AF361718

Analytical Technology in Nutrition Analysis

Analytical Technology in Nutrition Analysis

Special Issue Editor

Jose M. Miranda

MDPI • Basel • Beijing • Wuhan • Barcelona • Belgrade • Manchester • Tokyo • Cluj • Tianjin

Special Issue Editor
Jose M. Miranda
Universidade de Santiago de Compostela
Spain

Editorial Office
MDPI
St. Alban-Anlage 66
4052 Basel, Switzerland

This is a reprint of articles from the Special Issue published online in the open access journal *Molecules* (ISSN 1420-3049) (available at: https://www.mdpi.com/si/molecules/Nutrition_analysis).

For citation purposes, cite each article independently as indicated on the article page online and as indicated below:

LastName, A.A.; LastName, B.B.; LastName, C.C. Article Title. *Journal Name* **Year**, *Article Number*, Page Range.

ISBN 978-3-03928-764-2 (Hbk)
ISBN 978-3-03928-765-9 (PDF)

Contents

About the Special Issue Editor

Jose M. Miranda (Ph.D.) Prior to his academic and scientific activities, he worked for several years as a health inspector. Currently, he is working as assistant professor in the University of Santiago de Compostela (Spain). He has published over 100 scientific articles in indexed journals. His areas of expertise are food safety, food chemical analysis, characterization of compounds with nutritional interest, and metagenomics.

 *molecules*

Editorial

Analytical Technology in Nutrition Analysis

Jose M. Miranda

Laboratorio de Higiene Inspección y Control de Alimentos, Departamento de Química Analítica, Nutrición y Bromatología, Universidade de Santiago de Compostela, 27002 Lugo, Spain; josemanuel.miranda@usc.es; Tel.: +34-982-252-231 (ext. 22407); Fax: +34-982-254-592

Received: 9 March 2020; Accepted: 11 March 2020; Published: 17 March 2020

The great challenge facing humanity in the coming decades is to secure food for the 9.8 billion people who are expected to inhabit the planet by around 2050 and 11.2 billion in 2100 [1]. To increase the food production by traditional methods to meet this demand for food is very difficult. Additionally, traditional methods of food production, both of plant and animal origin, present specific problems that make it difficult for them to meet such an ambitious increase target.

In this sense, terrestrial agriculture presents a problem because fresh water (an essential resource) is an increasingly scarce commodity, and progressive desertification of the Earth's surface is taking place and will probably be aggravated in the future by global warming. About 45% of the world's land surface is currently considered drylands, while 12 million hectares of land are degraded yearly through a lack of water and related processes. According to the Food and Agriculture Organization of the United Nations [2], agricultural productivity is persistently declining at over 1% per year.

With respect to food production of animal origin, this also presents specific challenges, such as the fact that intensive production methods require large amounts of land, water and feed, and some livestock (such as ruminants) produce high levels of greenhouse gas emissions. Thus, intensive methods of animal production have serious drawbacks from the point of view of environmental care. In order to properly feed such a large population, it will be necessary to increase food production while respecting ecosystems and natural resources. The current high demand for animal proteins requires that livestock is reared in large numbers over diminishing land resource which is not possible and, therefore, alternative substitutes for animal proteins needs to be embraced to overcome this problem [3].

The abovementioned fact means that demand for food produced from non-traditional sources is expected to rise in the coming decade [4]. Fortunately, nowadays increasing acceptance for novel foods is also being observed, not only in developing but also in developed countries, which is mainly influenced by consumer awareness of the nutritional benefits linked to these kinds of foods [3]. As it can be seen in Figure 1, the investigation about novel foods has experienced a dramatic increase in the last decade (about three-fold).

In addition to the need for an increase in food production, nowadays, in most countries of the world, there is a growing prevalence of chronic non-communicable diseases, many of which are diet-related [5]. As a result, there is widespread consumer demand for foods with a nutritional composition more in line with current nutritional guidelines, and which include a greater proportion of the nutrients that have a potential beneficial effect on human health, or fewer of those components that have a negative effect on human health [6]. Therefore, both because of the need to ensure food safety in food, especially in those that do not have a history of safe use. In addition, there is also a need for analytical methodologies to reliably determine both the presence of specific nutrition-related components in foods, and the effects of these dietary components on human health, in areas as diverse as lipidomics, proteomics, transcriptomics, genomics, epigenomics, or metagenomics [7,8].

To meet these needs, it is essential that we in the international scientific community work intensively to ensure safe, effective and honest food production and to protect the health of consumers.

For this reason, from Molecules it was recognized the need to propose the Special Issue "Analytical Technologies in Nutrition Analysis". This Special Issue was aimed to offer an appropriate opportunity to all the contributors to make their results and techniques more visible, and to present the most recent findings.

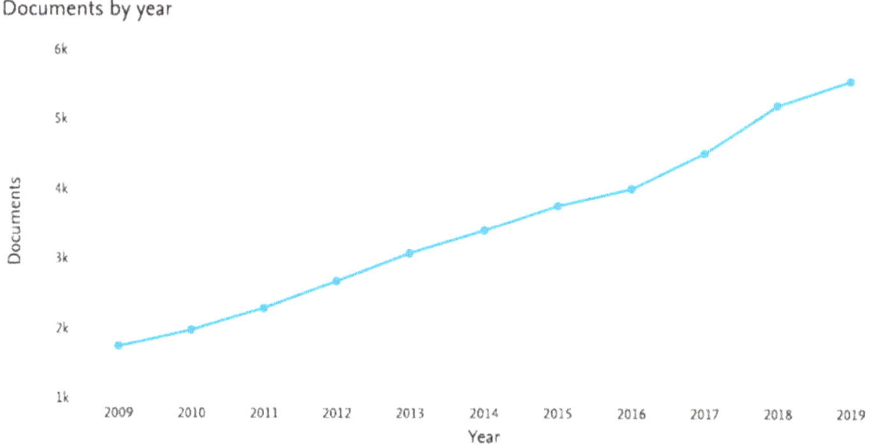

Figure 1. Results analysis for Scopus query "novel foods" in title, keywords or abstract section of the articles between 2009 and 2019.

This Special Issue has received remarkably positive feedback, with many contributions submitted by numerous geographically diverse scientists, resulting in a collection of 10 publications, including two exhaustive review articles [9–18]. Among the contributing authors, authors can be found from Asia (China, India, and Uzbekistan), Europe (Spain), South America (Brazil and Chile) and North America (Mexico). The published articles include findings related to the comprehensive bioactive compound profile and antioxidant capacities of mamey apple (*Mammea americana*), camapu (*Physalis angulata*), and uxi (*Endopleura uchi*) that can contribute to their economic exploitation [9].

Another article described the multifunctional activity of amaranth (*Amaranthus hypochondriacus* spp.) proteins, opening the possibility that amaranth hydrolyzed with alcalase and flavourzyme to be used as a value-added ingredient with multi-functional bioactive properties [10]. Another article aimed to find an efficient extraction method and investigate some of physical and chemical parameters, like water solubility, emulsification, foaming properties, and oil-holding capacity of obtained scorpion proteins. The results obtained suggest that scorpion proteins can be considered as an important ingredient and raw material for the creation of water-soluble supramolecular complexes for drugs [11].

A fractional factorial design was used to evaluate the effects of temperature, frying time, blanching treatment and the thickness of potato slices on a very relevant potential toxic compound (acrylamide) content in crisps. The findings obtained demonstrate that acrylamide concentration remained at 70% in fried chips, and reductions took place, mainly at the intestinal phase, as a result of reaction with nucleophilic compounds [12]. *N*-carbamylglutamate, a synthetic analogue of *N*-acetylglutamate, is an activator of blood ammonia conversion and endogenous arginine synthesis. This study will provide a solid foundation for the evaluation of availability and metabolic mechanism of *N*-carbamylglutamate in animals [13].

The *Artemisia argyi* leaf has been used as a traditional medicine and food supplement in Asian countries for hundreds of years. Phytochemical studies disclosed that *Artemisia argyi* leaf contains various bioactive constituents, mainly phenolic acids, which have great potential as possible alternatives to those organic solvents in health-related areas such as food and pharmaceuticals [14]. Regarding the use of seaweeds as alternative dietary fibre sources to terrestrial vegetables, in this Special Issue an article is presented evaluating the nutritional composition and physicochemical properties of two dried commercially interesting edible red seaweeds, *Gracilaria corticata* and *G. edulis*. In view of the results, both *G. corticata* and *G. edulis* contain important nutrients for human health and are possible natural functional foods [15]. More generally, a wide review about the current knowledge surrounding the impacts of seaweeds and their derived polysaccharides on the human microbiota is also presented, in which potential benefits against chronic non-transmissible diseases were discussed [16].

Finally, two articles describing potentially beneficial food sources of fat for humans are presented. In one on them, it was concluded that refined commercial salmon oil can be transformed into a profitable source of eicosapentaenoic and docosapentaenoic acids, thus leading to a product with higher commercial value, and that this process can be optimized by using response surface methodology [17]. The last article of the Special Issue consists of a review article about avocado oil, including discussion about the extraction methods, chemical composition, and various applications of avocado oil in the food and medicine industries. Based on the available data, avocado oil has established itself as an oil that has a very good nutritional value at low and high temperatures, with multiple technological applications that can be exploited for the benefit of its producers [18].

This Special Issue is accessible thought the following link: https://www.mdpi.com/journal/molecules/special_issues/Nutrition_analysis.

As Guest Editor for this Special Issue, I would like to thank all the authors and co-authors for their contributions and all the reviewers for their effort in carefully and rapid evaluating the manuscripts. Last but not least, I would like to appreciate the hard work done by the editorial office of the Molecules journal, as well as their kind assistance in preparing this Special Issue.

Funding: This research received no external funds.

Conflicts of Interest: The author declares no conflict of interest.

References

1. United Nations, Department of Economics and Social affairs, Population Division. *World Population Prospects 2019: Highlights*; United Nations: New York, NY, USA, 2019.
2. FAO. *The Future of Food and Agriculture–Trends and Challenges*; Food and Agriculture Organization of the United Nations: Rome, Italy, 2017.
3. Imathiu, S. Benefits and food safety concerns associated with consumption of edible insects. *NFS J.* **2020**, *18*, 1–11. [CrossRef]
4. Selenius, O.; Korpela, J.; Salminen, S.; Gallego, C.G. Effect of chitin and chitooligosaccharide on in vitro growth of *Lactobacillus rhamnosus* GG and *Escherichia coli* TG. *Appl. Food Biotechnol.* **2018**, *5*, 163–172.
5. Miranda, J.M.; Anton, X.; Redondo-Valbuena, C.; Roca-Saavedra, P.; Rodriguez, J.A.; Lamas, A.; Franco, C.M.; Cepeda, A. Egg and egg-derived foods: Effects on human health and use as functional foods. *Nutrients* **2015**, *7*, 706–709. [CrossRef] [PubMed]
6. Roca-Saavedra, P.; Mendez-Vilabrille, V.; Miranda, J.M.; Nebot, C.; Cardelle-Cobas, A.; Franco, C.M.; Cepeda, A. Food additives, contaminants and other minor components: Effects on human gut microbiota—A review. *J. Physiol. Biochem.* **2018**, *74*, 69–83. [CrossRef] [PubMed]
7. Lamas, A.; Regal, P.; Vázquez, B.I.; Miranda, J.M.; Franco, C.M.; Cepeda, A. Transciptomics: A powerful tool to evaluate the behavior of foodborne pathogens in the food production chain. *Food Res. Int.* **2019**, *15*, 108543. [CrossRef] [PubMed]
8. Ramon Vidal, D. Model organisms and "OMIC" technologies: New tools for the development of healthy foods. *Curr. Opin. Biotechnol.* **2013**, *24*, S19. [CrossRef]
9. Barbosa Lima, L.G.; Montenegro, J.; de Abreu, J.P.; Barros santos, M.C.; Pimenta do Nascimento, T.; da Silva santos, M.; Ferreira, A.G.; Cameron, L.C.; Larraz Ferreira, M.S.; Teodoro, A.J. Metabolite Profiling by UPLC-MSE, NMR, and Antioxidant Properties of Amazonian Fruits: Mamey Apple (*Mammea Americana*), Camapu (*Physalis Angulata*), and Uxi (*Endopleura Uchi*). *Molecules* **2020**, *25*, 342. [CrossRef] [PubMed]

10. Ayala-Niño, A.; Rodríguez-Serrano, G.M.; González-Olivares, L.G.; Contreras-López, E.; Regal-López, P.; Cepeda-Sáez, A. Sequence identification of bioactive peptides from Amaranth seed proteins (*Amaranthus hypochondriacus* spp.). *Molecules* **2019**, *24*, 3033. [CrossRef] [PubMed]

11. Wali, A.; Wubulikasimu, A.; Mirzaakhmedov, S.; gao, Y.; Omar, A.; Arken, A.; Yili, A.; Aisa, H.A. Optimization of Scorpion protein extraction and characterization of the proteins' functional properties. *Molecules* **2019**, *24*, 4103. [CrossRef] [PubMed]

12. Martinez, E.; Rodriguez, J.A.; Mondragon, A.C.; Lorenzo, J.M.; Santos, E.M. Influence of potato crisps processing parameters on acrylamide formation and bioaccesibility. *Molecules* **2019**, *24*, 3827. [CrossRef] [PubMed]

13. Ma, Y.; Zeng, Z.; Kong, L.; Chen, Y.; He, P. Determination of *N*-carbamylglutamate in feeds and animal products by high performance liquid chromatography tandem mass spectrometry. *Molecules* **2019**, *24*, 3172. [CrossRef] [PubMed]

14. Duan, L.; Zhang, C.; Zhang, C.; Xue, Z.; Zheng, Y.; Guo, L. Green extraction of phenolic acids from *Artemisia argyi* leaves by Tailor-made ternary deep eutectic solvents. *Molecules* **2019**, *24*, 2842. [CrossRef] [PubMed]

15. Rosemary, T.; Arulkumar, A.; Paramasivam, S.; Mondragon-Portocarrero, A.; Miranda, J.M. Biochemical micronutrient and physicochemical properties of the dried red seaweeds *Gracilaria edulis* and *Gracilaria corticata*. *Molecules* **2019**, *24*, 2225. [CrossRef] [PubMed]

16. Lopez-Santamarina, A.; Miranda, J.M.; Mondraon, A.C.; Lamas, A.; Cardelle-Cobas, A.; Franco, C.M.; Cepeda, A. Potential use of marine seaweeds as prebiotics: A review. *Molecules* **2020**, *25*, 1004. [CrossRef] [PubMed]

17. Dovale-Rosabal, G.; Rodriguez, A.; Contreras, E.; Ortiz-Viedma, J.; Muñoz, M.; Trigo, M.; Aubourg, S.P.; Espinosa, A. Concentration of EPA and DHA acid adduction reaction conditions using response surface methodology. *Molecules* **2019**, *24*, 1642. [CrossRef] [PubMed]

18. Flores, M.; saravia, C.; Vergara, C.E.; Avila, F.; Valdés, H.; Ortiz-Viedma, J. Avocado oil: Characteristics, properties, and applications. *Molecules* **2019**, *24*, 2172. [CrossRef] [PubMed]

Article

Metabolite Profiling by UPLC-MS^E, NMR, and Antioxidant Properties of Amazonian Fruits: Mamey Apple (Mammea Americana), Camapu (Physalis Angulata), and Uxi (Endopleura Uchi)

Larissa Gabrielly Barbosa Lima [1], **Julia Montenegro** [1], **Joel Pimentel de Abreu** [1],
Millena Cristina Barros Santos [2,3], **Talita Pimenta do Nascimento** [2,3], **Maiara da Silva Santos** [4],
Antônio Gilberto Ferreira [5], **Luiz Claudio Cameron** [3], **Mariana Simões Larraz Ferreira** [2,3]
and **Anderson Junger Teodoro** [1,*]

[1] Laboratory of Functional Foods, Nutrition Biochemistry Core, Food and Nutrition Graduate Program, Federal University of the State of Rio de Janeiro, UNIRIO. Av. Pasteur, 296, Rio de Janeiro 22290-240, Brazil; larissagabrielly_lima@hotmail.com (L.G.B.L.); juliamontenegro95@gmail.com (J.M.); pimenabreu@gmail.com (J.P.d.A.)

[2] Laboratory of Bioactives, Nutrition Biochemistry Core, Food and Nutrition Graduate Program, UNIRIO. Av. Pasteur, 296, Rio de Janeiro 22290-240, Brazil; barrosmillena@gmail.com (M.C.B.S.); talitapiment@gmail.com (T.P.d.N.); mariana.ferreira@unirio.br (M.S.L.F.)

[3] Center of Innovation in Mass Spectrometry, Laboratory of Protein Biochemistry, UNIRIO. Av. Pasteur, 296, Rio de Janeiro 22290-240, Brazil; cameron@unirio.br

[4] Fluminense Federal Institute of Education, Science and Technology, IFF, Av. Dário Viêira Borges, 235-Lia Márcia, Bom Jesus do Itabapoana, Rio de Janeiro 28360-000, Brazil; maiarasantos@yahoo.com.br

[5] Laboratory of NMR, Department of Chemistry, Federal University of São Carlos, UFSCar. Washington Luiz, s/n, São Carlos 13565-905, SP, Brazil; giba_04@yahoo.com.br

* Correspondence: atteodoro@gmail.com; Tel.: +55-21-25427236; Fax: +55-21-25427752

Received: 1 October 2019; Accepted: 12 December 2019; Published: 15 January 2020

Abstract: The metabolite profiling associated with the antioxidant potential of Amazonian fruits represents an important step to the bioactive compound's characterization due to the large biodiversity in this region. The comprehensive bioactive compounds profile and antioxidant capacities of mamey apple (*Mammea americana*), camapu (*Physalis angulata*), and uxi (*Endopleura uchi*) was determined for the first time. Bioactive compounds were characterized by ultra-performance liquid chromatography coupled to high resolution mass spectrometry (UPLC-MS^E) in aqueous and ethanolic extracts. Globally, a total of 293 metabolites were tentatively identified in mamey apple, campau, and uxi extracts. The main classes of compounds in the three species were terpenoids (61), phenolic acids (58), and flavonoids (53). Ethanolic extracts of fruits showed higher antioxidant activity and total ion abundance of bioactive compounds than aqueous. Uxi had the highest values of phenolic content (701.84 mg GAE/100 g), ABTS (1602.7 µmol Trolox g^{-1}), and ORAC (15.04 µmol Trolox g^{-1}). Mamey apple had the highest results for DPPH (1168.42 µmol TE g^{-1}) and FRAP (1381.13 µmol FSE g^{-1}). Nuclear magnetic resonance (NMR) spectroscopy results showed that sugars and lipids were the substances with the highest amounts in mamey apple and camapu. Data referring to chemical characteristics and antioxidant capacity of these fruits can contribute to their economic exploitation.

Keywords: Amazonian fruits; antioxidant; phenolic compounds; UPLC-MS^E; bioactive compounds

---

1. Introduction

The Amazonian region offers a wide variety of native fruits and most of them are typically obtained from nature or grown only for the local market supply in form of pulp or in natura. The economic exploitation is potentially of great importance for the region [1]. These native fruits have been studied as potential bioactive sources and many of them have shown high antioxidant capacity and elevated phenolic compounds content [2].

The antioxidant properties of Amazonian fruits have been the object of many researches, mainly due to the presence of natural antioxidants such as carotenoids and phenolic compounds. The group of antioxidant compounds found in fruits produced in Amazonian can protect the human body against toxic effects and preventing diseases such as chronic degenerative disorders, cardiovascular diseases, premature ageing, diabetes, and neurodegenerative diseases [3–5].

The Brazilian Amazonian region has a great biodiversity with approximately about 220 edible plant species producing fruit, representing 44% of native fruit diversity in Brazil. Some species such as mamey apple (*Mammea americana*), camapu (*Physalis angulata*), and uxi (*Endopleura uchi*) have widely appreciated flavors by Brazilian consumers, but still of moderate importance to the economy. They show potential for commercialization in both domestic and international markets [6].

Mamey apple (*Mammea americana*) is a fruit with a reddish yellow color, aromatic, and edible pulp, and is popularly known as abricó-do-pará, abricó, mammey apple or apricot from St. Domingo. This fruit has an important agroindustrial potential and this pulp can be used to produce different products such as syrup, juice, sherbet, jam, and pastes [7]. *Physalis angulata* L. is an herb indigenous of the *Solanaceae* family, and dispersed throughout tropical areas, including the Amazonian region [8]. This is popularly known as "camapu" and its juice is used as sedative, depurative, anti-rheumatic, and for the relief of earache, and is also used as a traditional medicine [9]. Uxi (*Endopleura uxi*) is an important fruit distributed in Pará and Amazonas, states of Northern Brazil. The only edible part of uxi presents a yellow-brownish color pulp with a rough-like texture, containing high content of fat (mainly oleic acid) and carotenoids, mostly trans-β-carotene, with an unique flavor [10].

Due to their peculiar biodiversity, knowledge of the species and functional property characterization of native Amazonian fruits such as mamey apple, camapu, and uxi present a major challenge to their appreciation, as many of these species are unexplored and their chemical properties remain unknown.

Few studies in the literature were found to provide a comprehensive metabolomic analysis of the bioactive compounds combined to antioxidant capacity of these fruits from the Amazonian biome. However, some studies have shown the metabolomic composition of Amazonian native fruits. Paz et al. [11] studied *Clavija lancifolia Desf.* using liquid chromatography mass spectrometry (LC–MS/MS) and found some compounds of flavonoids for which kaempferol was the main compound. Souza et al. [12] identified the phenolic compounds cyanidin 3-O-rutinoside, chlorogenic acid, and rutin in *Oenocarpus distichus* fruits, using high-performance liquid chromatography (HPLC). It has been reported the presence of catechin, caffeic acid, rutin, orientin, quercetin, apigenin, luteolin, and kaempferol in *Mauritia flexuosa* L. f. (*Arecaceae*), another fruit from Amazonian biome [13].

Concerning the importance of the characterization of bioactive compounds from Amazonian fruits, the aim of this study was to assess the phytochemical profile and the total antioxidant capacity of mamey apple (*Mammea americana*), camapu (*Physalis angulata*), and uxi (*Endopleura uchi*).

2. Results and Discussion

2.1. Total Phenolic Compounds Content

Mamey apple (MA), camapu (C), and uxi (U) were submitted for analysis of total phenolic compounds from extractions submitted with different vehicles solvents (water (W) and ethanol (E)) to determine the most efficient extraction solvent of the mentioned compounds. The results showed that the ethanolic extracts (MAE and UE) of the different Amazonian fruits showed a greater quantity of phenolic compounds when compared to the aqueous extracts (MAW and UW) (Figure 1). The preparation and extraction from this wide range of samples depends mostly on the nature of the sample matrix and the chemical properties of the phenolics, including molecular structure, polarity, concentration, number of aromatic rings, and hydroxyl groups. Conventional solid-liquid using organic solvent extraction is the main method used to extract phenolics. The sample preparation, polarity of the solvent used, the technique employed and temperature are factors that can influence the extraction and contents of these compounds [14].

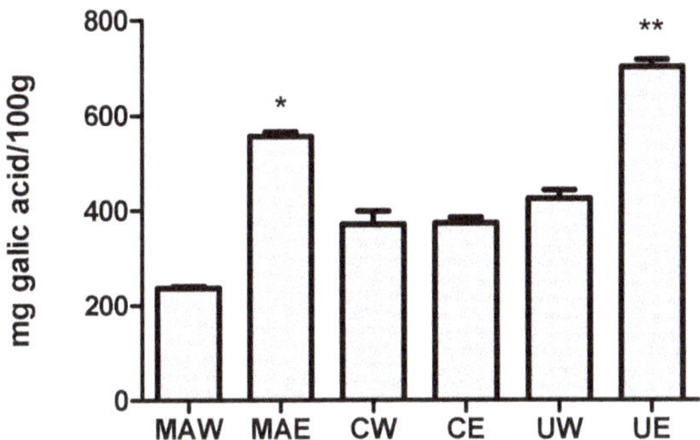

Figure 1. Total phenolic content in aqueous mamey apple (MAW), ethanolic mamey apple (MAE), aqueous camapu (CW), ethanolic camapu (CE), aqueous uxi (UW), and ethanolic uxi (UE) extracts. Results are expressed by mean ± SD (*n* = 3) and were compared by the one-way ANOVA test with post-test Tukey's (* $p < 0.01$; ** $p < 0.001$).

Uxi ethanolic extract (UE) showed the highest phenolic compound content (701.839 mg GAE/100 g), followed by mamey apple (MAE) (556.105 mg GAE/100 g), and camapu (CE) (237.39 mg GAE/100 g). The aqueous extracts of mamey apple (MAW) and uxi (UW) showed significant difference ($p < 0.05$) when compared to the respective ethanolic extracts. No significant differences ($p > 0.05$) were observed in the content of phenolic compounds between the aqueous and ethanolic extracts in the camapu sample.

Aqueous and ethanolic extract showed similar fruits yield extracts, being dependent on the moisture percentage of each fruit. Uxi yield was significantly higher (UE-28.82% and UW-29.42%), followed by mamey apple (MAE-14.04% and MAW-13.13%) and camapu (CE-10.50% and CW-11.10%). Considering these results, the levels of phenolic compounds per mass of fruits for the highest yield would be 202.27, 78.08, and 24.92 mg GAE/100 g for uxi, mamey apple, and camapu, respectively.

Péroumal et al. [15] studied the pulp of six mamey apple accessions and found values for total phenolic content between 90 and 143 mg GAE/100 g. It can be observed that the results obtained for uxi ethanolic extracts in this work are higher than some Amazonian fruits such as araçá-boi (87.2 ± 3.0 mg GAE/100 g) and araçá (129.1 ± 9.3 mg GAE/100 g), and lower than camu-camu (1797.2 ± 37.7 mg GAE/100 g) reported by Genovese et al. [16]. Comparing the results of this study with fruits from the Brazilian Cerrado biome, the Amazonian fruits present lower phenolic compounds in relation to sweet passion fruit (245.36 ± 3.70 mg GAE/100 g), soursop (281 ± 5.40 mg GAE/100 g), murici (334.37 ± 9.07 mg GAE/100 g), and marolo (739.37 ± 7.92 mg GAE/100 g) [17]. According to the classification proposed by Vasco et al. [18], MAE, uxi extracts were classified as fruits as having with medium phenolic content (100–500 mg GAE/100 g).

2.2. Phytochemical Profile by UPLC-MS^E

For the first time, bioactive compounds of Amazonian fruits like mamey apple, camapu, and uxi were elucidated by UPLC-MS^E metabolomic approach. Globally, 293 compounds were tentatively identified from aqueous and ethanolic extracts of these fruits (Table S1) and relatively quantified taking account all extracts based on ion counting. Table 1 shows the number of identified compounds and their classification into eight chemical classes according to Phenol Explorer database [19]: phenolic acid, flavonoids, chalcones, coumarins, amino acid related compounds, fatty acid, and terpene related compounds. The main bioactive compounds in the three species were terpenoids (*n* = 61; 21%), phenolic acids (n = 58; 20%), and flavonoids (n = 53; 18%). Other metabolites were also identified such as other polyphenols (n = 50; 17%), including lignans, coumarins and tannins, and also other metabolites (n = 56; 19%) such as amino acid related, alkaloids and polyketides, showing the extraction and LC-MS methods were suitable to characterize different polarities compounds.

Table 1. Number of bioactive compounds distributed by classes and other compounds identified in aqueous (W) and ethanolic (E) extracts of mamey apple, camapu, and uxi.

Compounds (%)	Mamey Apple		Camapu		Uxi	
	W	E	W	E	W	E
Phenolic Acid	43 (20.57%)	37 (19.68%)	44 (24.04%)	40 (21.51%)	38 (23.17%)	35 (21.21%)
Flavonoids	36 (17.22%)	30 (15.96%)	33 (18.03%)	29 (15.59%)	24 (14.63%)	27 (16.36%)
Chalcones	2 (0.96%)	2 (1.06%)	2 (1.09%)	2 (1.08%)	3 (1.83%)	3 (1.82%)
Coumarins	16 (7.66%)	15 (7.98%)	14 (7.65%)	14 (7.53%)	8 (4.88)	9 (5.45%)
Others phenolic compounds	27 (12.92%)	25 (13.30%)	23 (12.57%)	30 (16.13%)	20 (12.20%)	26 (15.76%)
Amino acid related compounds	17 (8.13%)	14 (7.45%)	17 (9.29%)	20 (10.75%)	18 (10.98%)	15 (9.09%)
Fatty acids related compounds	22 (10.53%)	22 (11.70%)	11 (6.01%)	12 (6.45%)	14 (8.54%)	15 (9.09%)
Terpenoids	46 (22.01%)	43 (22.87%)	39 (21.31%)	39 (20.97%)	39 (23.78%)	35 (21.21%)
Total of compounds	209 (100%)	188 (100%)	183 (100%)	186 (100%)	164 (100%)	165 (100%)

The identification of different phenolic compounds in this study makes it relevant, since the presence of these compounds has a range of bioactivities, as already reported in vitro and in vivo studies [20]. In a recent study, extracts of phenolic compounds from jatobá-do-cerrado can inhibit α-amylase and α-glycosidase after in vitro digestion and modulate the glucose metabolism [21].

Mamey apple was characterized by high number of bioactive compounds. Among the identified metabolites, 209 (71%) and 188 (64%) were found in aqueous and ethanolic extracts of mamey apple, respectively. Furthermore, 183 (62%) and 186 (63%) compounds were found in the aqueous extract ethanolic extracts of camapu, respectively. In uxi, 164 (56%) and 165 (56%) compounds were tentatively identified in the aqueous and ethanolic extracts, respectively. Notably, 50 (17%) compounds were found only in mamey apple (e.g., droserone, norlichexanthone, procyanidin C1, 3,4-leucopelargonidin II, and piquerol A). Piquerol A is a sesquiterpene with low stability in nature and has only previously been tested as an insecticide and as an inhibitor of metabolism in cell cultures [22]. About 43 (15%) were tentatively identified only in camapu (e.g., myristicin, 11-deoxocucurbitacin, synapic acid I, and pseudopurpurin). Medina et al. [23] reported the presence of sinapic acid in *Passiflora edulis Sims*. Furthermore, about 36 (12%) compounds were found exclusively in the uxi extract (e.g., 4-coumaroylchiquimate, ferulic acid I, kaempferol II, jacareubin, 6-deoxyjacareubin, vanylactic acid II, nigakilactone A, and jasmonic acid). Kaempferol is an important flavonol identified in other fruits like hybrid grapes by Rosso et al. [24].

These metabolites are antioxidant inhibitors of mutagenic and carcinogenic compounds and are considered neuroprotective agents in neurodegenerative disorders, such as Parkinson's and Alzheimer's diseases [25].

The acid phenolic profiling was the most abundant subclass of phenolic compounds in all extracts evaluated, as shown in the Tables 1 and 2. Aqueous and ethanolic camapu extracts (CW and CE) showed about 24 and 22% of phenolic acids, respectively, followed by mamey apple aqueous (MAW) and ethanolic (MAE) extracts. In addition to phenolic acids, flavonoids showed high number of identification as well as elevated abundance relative.

Table 2. Flavonoids tentatively identified in mamey apple, camapu, and uxi extracts.

Possible Identifications	CAS	*m/z* Exp	RT (min)	Fragment *m/z*	Error (ppm)	Mamey		Camapu		Uxi	
						W	E	W	E	W	E
Genistin	529-59-9	431.0978	4.55	**133.0294 (10.20)**	−1.23	–	–	–	–	X	X
Eriodictiol I	552-58-9	287.0552	3.05	81.0338 (5.46); 93.0339 (78.57); 119.0496 (100); 155.0342 (10.19); 163.0395 (60.20)	−2.86	X	X	–	–	–	–
Hesperidin	520-26-3	609.1875	0.43	79.0188 (0.66); 369.0671 (50.27); 488.1618 (10.04)	8.37	X	X	X	X	X	X
Narirutin	14259-46-2	579.1775	0.43	72.9924 (78.62); 79.0188 (0.58); 117.0187 (26.50); 135.0294 (87.65); 357.1033 (3.92); 369.0671 (44.06); 535.1514 (21.08)	9.66	X	X	X	X	X	X
Pomiferin I	572-03-2	419.1505	2.62	nd	1.21	X	X	X	X	X	X
Ononin	486-62-4	429.1170	2.02	nd	−4.89	X	X	X	X		
Mammeisin	18483-64-2	405.1677	4.01	109.0653 (0.46); 154.0615 (0.72)	−7.30	X	X				
Quercitrin	522-12-3	447.0923	3.48	151.0030 (2.20); 285.0393 (5.36)	−2.09	X	X	X			
Quercetin 3-galactoside	482-36-0	463.0875	3.47	151.0030 (2.97); 255.0291 (5.58); 271.0241 (16.99); 285.0393 (7.25); 300.0266 (31.67)	−1.44	X	X	X	X	X	X
Kaempferol I	520-18-3	285.0396	4.90	nd	−2.70	X	X				X
Dihydroquercetin	480-18-2	303.0505	1.42	147.0120 (100)	−1.65	X	X				
Luteoforol	24897-98-1	289.0733	0.56	109.0289 (12.30)	5.54	X	X	X	X	X	X

Terpenes were between the most abundant bioactive classes in all extracts. About 46 and 43 terpenes, representing 22% of identifications, were found in the MAW and MAE extracts, respectively. In UE extracts, 39 (23%) terpenes were tentatively identified too. Terpenoids are important secondary metabolites of plants and are extremely chemically diversified, being estimated at more than 40,000 substances. Its use has been described as flavoring agents, in addition to providing benefits to human health through its antioxidant potential [26,27].

Although the number of identifications between the extracts of the same fruit was not significantly different, the ethanolic extracts showed a greater relative abundance of compounds in relation to the aqueous extracts in all fruits studied (Figure 2). This fact can be explained by the large variability of the structures of these bioactive compounds and the proportion of organic solvents with water is required for better extraction.

The principal component analysis (PCA) and hierarchical cluster analysis (HCA) were applied to explain possible difference among extracts (Figures 3 and 4). First, the PCA biplot represented both loading (metabolites) and scores (extracts). Such parameters distinguished the profile of bioactive compounds among the fruits evaluated in all extracts. Principal components (PC1 and PC2) demonstrated that the fruits extracts have different composition of bioactive compounds, but the use of aqueous and ethanolic extractors did not favor the removal of distinct metabolite profile in the species. The two major principal components (PC1 and PC2) explained more than 72% of the variance pattern (Figure 3). Discriminatory metabolites, which showed maximum variance (eigenvalues) among extracts, were observed mainly in the PC2, including the following bioactive compounds bryophyllin A II, 1-*O*-2′-hydroxy-4′-methoxycinnamoyl-b-D-glucose I, mammeisin II, eriodictiol II, justicidin A, hallactone B II, 7,2′-dihydroxy-4′-methoxy-isoflavanol, eleganin I, 5-hydroxyferulic acid methyl ester II, lancerin II, 6-methoxytaxifolin II, zapoterin, salvinorin A, meconic acid, benzoic acid II, auriculoside, leucocyanidin II, lophophorine, lancerin I, vernodalol, visnagin, sinapic acid II, sinapyl alcohol II, 6-methoxytaxifolin I, isobrucein A, and syringin I.

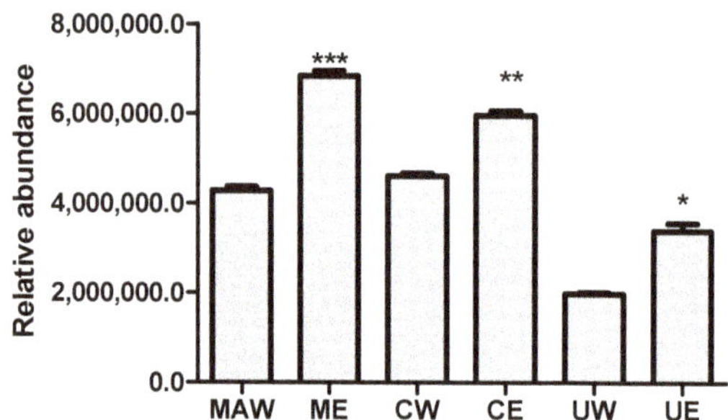

Figure 2. Relative abundance based on total ion counting of identified compounds from aqueous mamey apple (MAW), ethanolic mamey apple (MAE), aqueous camapu (CW), ethanolic camapu (CE), aqueous uxi (UW), and ethanolic uxi (UE) extracts. Results are expressed by mean ± SD ($n = 3$) and were compared by the one-way ANOVA test with post-test Tukey's (* $p < 0.05$; ** $p < 0.01$; *** $p < 0.001$).

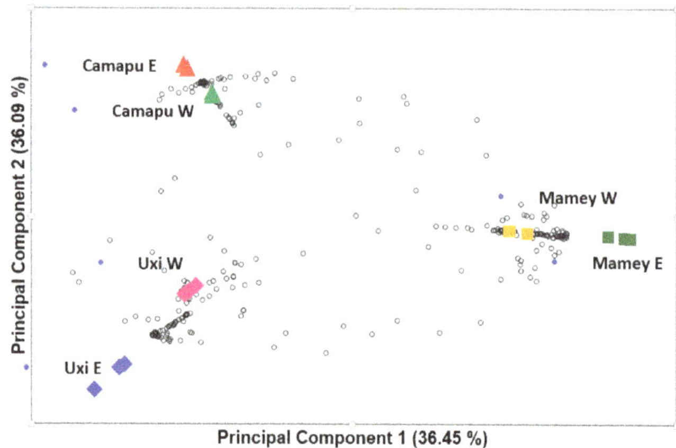

Figure 3. Principal component analysis (PCA) biplot (loadings and scores) of the bioactive compounds tentatively identified (loadings, empty circles) in the mamey (squares), camapu (triangles), and uxi (diamonds) fruits extracted with aqueous (W) and ethanolic (E) solvents.

Then, HCA was applied to observe the similarity/dissimilarity among the abundance of the discriminatory polyphenols (Figure 4) in order to understand the profile of composition among extracts. The heatmap indicated that metabolites (20) found in uxi extracts showed higher abundance in comparison to other extracts, especially in the ethanol extracts. Mamey also showed a distinguished bioactive profile, with six compounds found exclusively. Among them, sinapoyl alcohol II is an exclusive metabolite with high abundance in the mamey apple extracts. One of the metabolites with highest abundance in these extracts was syringin. This compound was determined and reported from fresh jaboticaba methanolic extract by Wu et al. [28] and implicated as immunomodulator having an anti-allergic effect [29].

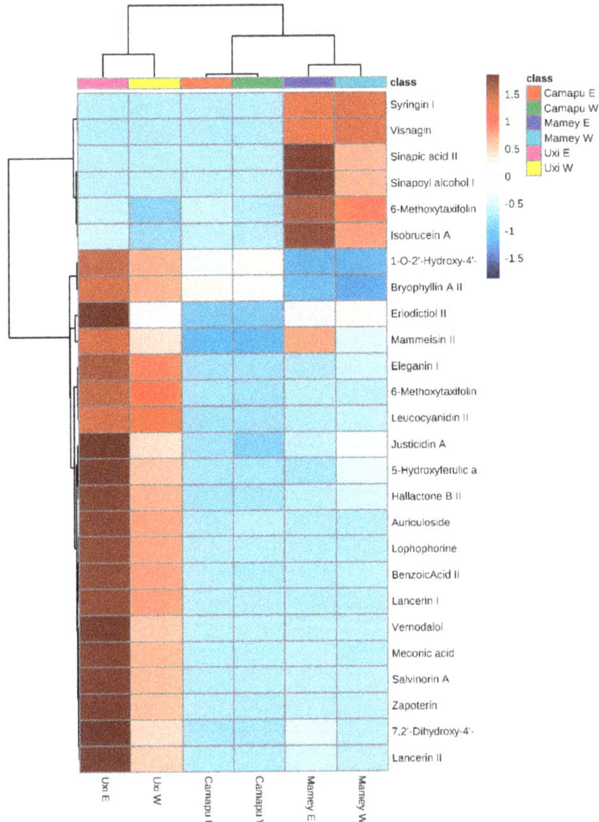

Figure 4. Hierarchical cluster analysis (HCA) and heatmap of the bioactive compounds in the mamey apple, camapu, and uxi fruits extracted with aqueous (W) and ethanolic (E) solvents, which showed maximum variance (eigenvectors) among extracts.

2.3. In Vitro Antioxidant Activity

Due to the multifunctional characteristics of the phenolic compounds found in Amazonian fruits, the effectiveness of measuring the antioxidant capacity of a pulp extract is better evaluated when using commonly accepted tests. The ethanolic and aqueous fruits extracts was determined by four different methods with different action mechanism (ABTS, DPPH, FRAP, and ORAC) (Table 3). According to Barros et al. [14], a single assay does not accurately account for all of the groups of antioxidant compounds, because of complexity of fruit matrices, and because of these methodologies can suffer interferences.

ABTS assay revealed values ranged between 263.67 ± 23.90 and 1602.7 ± 30 μmol Trolox g^{-1}. The highest antioxidant capacity was presented in UE samples (1602.7 ± 30.16 μmol Trolox g^{-1}). Freitas et al. [30] analyses fresh pulp of uxi and found 51.6 mg TE/100 g for ABTS. In general, the ethanolic extracts of fruits showed the higher antioxidant capacity for this methodology. However, the aqueous extract from camapu fruit was an exception with a value of 432.74 ± 16.17 μmol Trolox g^{-1} (Table 3). Schiassi et al. [31] studied the activity antioxidant of methanolic extracts of araça (10.92 ± 0.11 μmol Trolox g^{-1}), buriti (6.03 ± 0.00 μmol Trolox g^{-1}), cagaita (29.32 ± 0.69 μmol Trolox g^{-1}), yellow mombin (5.55 ± 0.01 μmol Trolox g^{-1}), and marolo (132.16 ± 1.40 μmol Trolox g^{-1}) and found values lower than the present study.

Table 3. Antioxidant activity of the mamey apple, camapu, and uxi samples by ABTS, DPPH, FRAP, and ORAC assays.

Assay	Mamey Apple		Camapu		Uxi	
	Aqueous	Ethanolic	Aqueous	Ethanolic	Aqueous	Ethanolic
ABTS (μmol Trolox g^{-1})	263.67 ± 23.90 [a]	937.66 ± 218.49 [b]	432.74 ± 16.17 [c]	419.43 ± 18.55 [c]	271.86 ± 22.14 [a]	1602.7 ± 30.16 [d]
DPPH (μmol Trolox g$^{-1)}$)	336.60 ± 3.05 [a]	1168.42 ± 218.56 [b]	386.24 ± 116.99 [c]	705.77 ± 100.74 [d]	46.95 ± 17.17 [e]	509.27 ± 26.95 [f]
FRAP (μmol ferrous sulphate g^{-1})	564.18 ± 18.90 [a]	1381.13 ± 189.95 [b]	970.60 ± 28.92 [c]	1183.98 [c] ± 46.62 [b]	376.66 ± 1.81 [d]	448.68 ± 41.97 [e]
ORAC (μmol Trolox g^{-1})	5.17 ± 0.56 [a]	8.88 ± 0.52 [b]	12.30 ± 1.15 [c]	11.15 ± 0.42 [c]	14.33 ± 1.36 [d]	15.04 ± 0.84 [d]

Results are expressed as mean ± standard deviation. Different letters on the same line show significant difference. Results were compared by the One-way ANOVA test with Tukey post-test ($p < 0.05$).

Antioxidant capacity was also evaluated by DPPH• radical scavenging method (Table 3) and expressed in aqueous and ethanolic extract concentration μmol Trolox g^{-1} sample. The highest activity was found in MAE (1168.42 μmol Trolox g^{-1}), followed by CE (705.771 μmol Trolox g^{-1}) and UE (509.68 μmol Trolox g^{-1}). According to Huchin et al. [32], the DPPH method can analyze hydrophilic and lipophilic compounds, and ethanol is one of the most used solvents for bioactive compounds extraction due, because proposes some benefits, such as low toxicity, good extraction produce, it is safe for human consumption and allows the extracts to be used in the food industry [33].

Antioxidant activity by FRAP method followed the same pattern of the DPPH analysis, where higher values were found for the ethanolic extracts. It is probable that the antioxidant compounds of the samples detected by FRAP are the same with those evaluated by DPPH. MAE also presented the highest activity (1381.13 μmol ferrous sulphate g^{-1}) These results was lower than obtained by Barros et al. [14] for aqueous extracts of achachairu (712.35 ± 6.61 μmol ferrous sulphate g^{-1}), araça-boi (798.92 ± 1.52 μmol ferrous sulphate g^{-1}), and bacaba (6567.45 ± 4.25 μmol ferrous sulphate g^{-1}).

Regarding to ORAC assay, the results revealed that UE and UW obtained the highest value (15.04 ± 0.84 and 14.33 ± 1.36 μmol Trolox g^{-1}), followed by CE and CW (11.15 ± 0.42 and 12.30 ± 1.15 μmol Trolox g^{-1}). No significant differences ($p > 0.05$) were observed between the aqueous and ethanolic extracts of uxi and camapu sample. MAE (8.88 ± 0.52 μmol Trolox g^{-1}) and MAW (5.17 ± 0.56 μmol Trolox g^{-1}) showed the lowest values ($p > 0.05$) between extracts. Santos et al. [34] reported to Amazonian fruits such as bacaca (195.00 ± 10.00 μmol Trolox g^{-1}), buriti (83.00 ± 6.00 μmol Trolox g^{-1}), inajá (26.00 ± 2.00 μmol Trolox g^{-1}), pupunha (94.00 ± 1.00 μmol Trolox g^{-1}), and tucumã (64.00 ± 4.00 μmol Trolox g^{-1}) results higher than camapu, mamey apple and uxi extracts.

2.4. Correlation between Total Phenolic Compounds and Antioxidant Capacity

The results of Pearson correlation coefficients (r) between total phenolic compounds and antioxidant capacity of aqueous and methanolic extracts suggest that the total phenolic compounds contributed to the in vitro antioxidant capacity of the extracts according to the method used. The aqueous extract showed a good association of total phenolic compounds with ABTS ($r = 0.843$) and ORAC ($r = 0.752$) for camapu fruit. ORAC assay has been largely applied to the assessment of free radical scavenging capacity of pure antioxidant compounds and antioxidant plant extracts [35]. The radical cation ABTS [2,29-]azinobis-(3-ethylbenzothiazoline-6-sulfonic acid) is one of the spectrophotometric methods used to measure water-soluble as well as lipid-soluble antioxidants, pure compounds, and food extracts. The pre-formed radical monocation of 2,29-azinobis-(3-ethylbenzothiazoline-6-sulfonic acid) (ABTS•^{1}) is generated by oxidation of ABTS with potassium persulfate and is reduced in the presence of such hydrogen-donating antioxidants [36]. 1-*O*-2′-Hydroxy-4′-methoxycinnamoyl-b-D-glucose, a compound present in CW (Figure 4), this compound belongs to the sub class of organic compounds known as hydroxycinnamic acid glycosides. Hydroxycinnamic acids are natural antioxidants found in fruits, vegetables, and cereals [37].

Regarding to mamey-apple, the ethanolic extract presented positive correlation with total phenolic compounds in relation to DPPH assay ($r = 0.910$). The 2,2-diphenylpicrylhydrazyl (DPPH) assay is widely used in plant biochemistry to evaluate their properties for scavenging free radicals. The method is based on the spectrophotometric measurement of DPPH concentration change resulting from the reaction with an antioxidant [38].

According to Figure 4, the acids syrigin I and visnagin were components with high relative abundancy. Visnagin is an antioxidant furanocoumarin derivative and is a furanochromone that is furo [3,2-g]chromen-5-one which is substituted at positions 4 and 7 by methoxy and methyl groups, respectively. Syringic acid is a dimethoxybenzene that is 3,5-dimethyl ether derivative of gallic acid. It has a role as a plant metabolite, a member of benzoic acids, and derives from a gallic acid. Previous reports described that hydroxybenzoic acids itself and its derivatives showed antioxidant properties against different type of free radicals and can prevent or decrease overproduction of reactive species. The main structural feature responsible for the antioxidative and free radical scavenging activity in the case of phenolic derivatives is the phenolic hydroxyl group. Phenols are able to donate the hydrogen atom of the phenolic OH to the free radicals, thus stopping the propagation chain during the oxidation process. This effect is modulated by the ring substituents, so that electron-withdrawing groups increase the bond-dissociation enthalpy, due to the stabilization of the phenol by a polar structure that leaves a positive charge on the OH group [39–41].

In uxi ethanolic extract, some bioactive compounds were identified with great relative abundance, such as auriculoside, that is a flavan glycoside. Most natural flavans are lipid-soluble and prominent in fruit skin or peel and leaf surfaces, and are usually found at higher concentrations in immature fruits compared to mature fruits [42]. Lancerin a plant metabolite found in UE is a member of xanthones, a C-glycosyl compound and a polyphenol. Among the polyphenols, the xanthones derivatives comprise an important class of the oxygenated heterocycles with a diversity of substitution patterns that have been described for their antioxidant activity, show to act as metal chela-tors, free radical scavengers, as well as inhibitors of lipid peroxidation [43]. The presence of these compounds may have contributed to the strong correlation ($r = 0.910$) found between total phenolic compounds and ORAC assay.

These results demonstrate the importance of do different methods to available the antioxidant activity, above all in complex matrices. The positive correlation between compound phenolics and ORAC, can be explained due to this assay may be considered a more exact method, because it uses a biologically relevant radical source (peroxyl) and allows the measurement of total antioxidant capacity through the combination of the antioxidant capacity of hydrophilic and lipophilic fractions [44]. According to the methodology used, these results suggest that phenolic compounds may be one of the main factors responsible for the antioxidant capacity of mamey apple, camapu, and uxi.

2.5. NMR Profile

Table 4 described the identified compounds for each fruit, with the chemical shift and signal multiplicity, as well as coupling constant (*J*), and the quantification in mg/g of freeze-dried fruit. Camapu had the greatest number of different compounds, indicating that it has a large variety of nutrients and metabolites, even more than could be extracted and identified. Uxi had the least amount, probably due to the great quantity of lipophilic compounds that could not be extracted, so this result does not mean that there are few compounds in uxi.

Sugars were most abundant compounds in those fruits. Camapu showed the highest amount of sucrose, and mamey apple showed the highest amount of fructose and glucose, both α and β. In the uxi extracts no sugars were found. This result indicates that mamey apple and camapu are sweeter than uxi, and probably would be more accepted by consumers. The sugars present in mamey apple were sucrose (1.50 ± 0.10 mg/g), fructose (0.85 ± 0.03 mg/g), α-glucose, and β-glucose (0.49 ± 0.02 and 0.40 ± 0.01 mg/g, respectively). The sugars found in camapu were sucrose (4.09 ± 0.46 mg/g), followed by fructose, α-glucose, and β-glucose (0.11 ± 0.00, 0.04 ± 0.00, and 0.04 ± 0.00 mg/g respectively).

Fruit and vegetable flavor depend upon taste (given by sweetness, sourness, acidity, and astringency) and aroma (concentrations of odor-active volatile compounds). Sweetness is determined by the concentrations of the predominant sugars, which are ranked relative to sucrose (fructose > sucrose > glucose). Sourness or acidity is determined by the predominant organic acids, which are ranked relative to citric acid (citric > malic > tartaric). Some amino acids, such as aspartic and glutamic, may also contribute to sourness. In general, consumer acceptance is related to soluble solids concentration (sugars, organic acids, soluble pectin, some phenolic compounds, and ascorbic acid) or the ratio of soluble solids to titratable acidity [45].

Table 4. NMR fingerprinting identification and quantification of mamey apple, camapu, and uxi.

Fruit	Compound	$\delta\,^1$H (ppm)	Multiplicity (J)	Mass (mg/g)
Mamey Apple	Formic acid	8.44	s	0.01 ± 0.00
	Shikimic acid	6.48	m	0.11 ± 0.01
	Sucrose	5.42	d (3.88)	1.51 ± 0.11
	α-glucose	5.22	d (3.76)	0.49 ± 0.02
	β-glucose	4.64	d (7.94)	0.40 ± 0.02
	Fructose	4.10	d (3.42)	0.85 ± 0.04
	Choline	3.19	s	0.01 ± 0.00
	Ethanol	1.18	t (7.09)	0.01 ± 0.00
Camapu	Formic acid	8.44	s	0.01 ± 0.00
	Sucrose	5.42	d (3.91)	4.09 ± 0.46
	α-glucose	5.23	d (3.73)	0.04 ± 0.01
	β-glucose	4.65	d (7.92)	0.04 ± 0.01
	Fructose	4.11	d (3.42)	0.12 ± 0.01
	Choline	3.20	s	0.01 ± 0.00
	Aspartic acid	2.82	dd (17.4; 3.78)	0.04 ± 0.01
	Acetic acid	1.94	S	0.01 ± 0.00
	GABA	1.89	quin (7.44)	0.02 ± 0.01
	Alanine	1.47	d (7.2)	0.02 ± 0.01
	Lactic acid	1.33	d (6.68)	0.01 ± 0.00
	Ethanol	1.18	t (7.09)	0.02 ± 0.001
	Valine	1.04	d (7.07)	0.01 ± 0.00
Uxi	Linoleic acid	5.32	M	1.79 ± 0.11
	Acetic acid	1.94	S	0.16 ± 0.01
	Alanine	1.45	d (7.20)	0.04 ± 0.01
	Ethanol	1.16	t (7.10)	0.15 ± 0.02
	Valine	1.05	d (7.05)	0.05 ± 0.001

It has been shown that mamay apple has ethanol (0.01 ± 0.00 mg/g), which is possibly a fruit sugar fermentation product. It was also found choline (0.01 ± 0.00 mg/g) and organic acids, such as formic acid and shikimic acid (0.01 ± 0.00 and 0.11 ± 0.00 mg/g) in mamey apple. Shikimic acid (3,4,5-trihydroxy-1-cyclohexene-1-carboxylic acid) is natural acid from the plant metabolism and common in berries fruits. It is a precursor for the biosynthesis of primary metabolites such as aromatic amino acids and folic acid, and a great many other aromatic compounds. The benzene ring is formed through the shikimate pathway, and shikimic acid is an extremely essential compound in plants and microbes. Shikimic acid has been found to occur in many tissues of a variety of plants, with a sufficiently high percentage. Moreover, its content and accumulation in different tissues depends on the rate of metabolic processes taking place in them. Shikimic acid is utilized for industrial synthesis of the oseltamivir (antiviral), zeylenone (employed as a preparation for chemotherapy of cancer), and monopalmityloxy shikimic acid (anticoagulant activity). In addition, shikimic acid derivatives represent a great interest for agriculture because many of them are used as herbicides and antibacterial agents [46].

Choline is a natural amine that can be synthesized in human body, but this production usually is not enough to meet human needs in men and postmenopausal women. Some important functions of choline are it is a part of the neurotransmitter acetylcholine; it is a part of the predominant phospholipids in membranes; it forms betaine, which is an important osmolyte in the kidney glomerulus and helps with the reabsorption of water from the kidney tubule. Eggs and liver are the main sources of choline, but many other foods contain significant amounts of choline and esters of choline [47].

Ethanol (0.02 ± 0.00 mg/g) was also present in camapu, indicating fermentation. The organic acids found in camapu were lactic acid, acetic acid and formic acid (0.01 ± 0.00, 0.01 ± 0.00, and 0.01 ± 0.00). Besides those organic acids, it was found γ-amino butyric acid (GABA) in camapu (0.02 ± 0.00 mg/g). GABA is a four-carbon nonprotein amino acid and is widespread in bacteria, animals, and plants. GABA is naturally present in small quantities in plants, and is produced in response to anaerobic conditions, γ-radiation, low pH, low or high temperatures, and darkness, and by mechanical manipulation. Some functions of GABA in plants are: regulation of cytosolic pH, protection against oxidative stress, defense against insects, and the regulation of pollen tube growth and guidance [48]. In vertebrates is a neurotransmitter that is deficient in the brain of people with Alzheimer disease [49]. Camapu has been shown to be effective in delaying the development of this disease and in its treatment, by increasing the proliferation of neural stem cells in vivo [50] and also shown an anxiolytic effect [51] which the authors suppose is due to the presence of GABA agonists. Several amino acids such as aspartic acid, alanine, and valine (0.03 ± 0.00, 0.02 ± 0.00, and 0.01 ± 0.08 mg/g) and choline (0.01 ± 0.00 mg/g) were also present in camapu in sufficient quantities to be identified.

For uxi, the compound with in the highest amount was linoleic acid (1.79 ± 0.10 mg/g), which is the main fatty acid of ω-6 group. This result shows that there are fatty acids that are sufficient to appear in a spectrum of fingerprints, even though other lipids may not have been completely extracted from the sample. There is probably a low amount of sugar in uxi fruit because there are high amounts of lipids and usually sugar and lipids in the pulp of fruits are inversely proportional. However, it was found ethanol (0.15 ± 0.01 mg/g), indicating that some sugar that was present before it was fermented. It was also found enough amount of acetic acid (0.15 ± 0.00 mg/g) valine and alanine amino acids (0.05 ± 0.00 and 0.04 ± 0.00 mg/g, respectively).

3. Material and Methods

3.1. Standards and Chemicals

2,2-Diphenyl-1-picrylhydrazyl (DPPH•), 2,2′-azino-bis (3-ethylbenzothiazoline-6-sulfonic acid) diammonium salt (ABTS), gallic acid, quercetin, 2′,7′-dichlorofluorescin diacetate (DCFH-DA), L-(+)-ascorbic acid (AA). Phenolic standards (caffeic acid, (+)-catechin, ellagic acid, (−)-epicatechin, gallic acid, gentisic acid, 4-hydroxybenzoic acid, myricetin, pyrogallol, quercetin, and quercetin 3-O-glucoside) were purchased from Sigma-Aldrich (St. Louis, MO, USA). All the solvents used for the UPLC-MS^E analysis were of LC-MS purity grade and were also purchased from Sigma-Aldrich (St. Louis, MO, USA).

3.2. Samples

The fruits in natura of mamey apple (*Mammea americana*), camapu (*Physalis angulata*), and uxi (*Endopleura uchi*) were obtained from producing regions of Pará State, in the months of January and February 2017. They were transported in vacuum plastic containers to Federal University of the State of Rio de Janeiro (UNIRIO, RJ, Brazil) and stored at −18 °C. The part of the fruit used for extraction was carried out according to their consumption, for mamey apple only pulp, for camapu the whole fruit and for uxi pulp and bark.

3.3. Sample Preparation

Samples were extracted by means of 2 extraction solvents: (I) ethanol 70% and (II) water, from 5 g of sample and 100 mL of extraction solvent, followed by homogenization at room temperature (−20 °C) in a homogenizer TE-420 (Tecnal, São Paulo, Brazil) for 10 min in the absence of light. The samples were then centrifuged (Thermo Fisher Scientific, Bartlesville, CA, USA) (5000× g, 5 min, 20 °C) and filtered through analytical filter paper. The extraction solvents were evaporated using a rotary evaporator under vacuum (Savant, Thermo Scientific) and the extracts were then frozen at −80 °C in an ultra-freezer and lyophilized (Terroni® LD 300, São Carlos, SP, Brazil) for 24 h. After this process, extracts were frozen at −20 °C until use in the experiments.

For UPLC-MS^E analysis, the extraction solvents were evaporated using a rotary evaporator under vacuum (Savant, Thermo Scientific, USA) at 40 °C and thereafter ressuspended in methanol/acetonitrile/water (2:5:93; *v/v*). Stock solutions (500 ppm) of 11 standards (caffeic acid, (+)-catechin, ellagic acid, (−)-epicatechin, gallic acid, gentisic acid, 4-hydroxybenzoic acid, myricetin, pyrogallol, quercetin and quercetin 3-*O*-glucoside from Sigma-Aldrich were prepared individually by dissolving accurately weighed amounts of standards in aqueous methanol an aliquot of each stock solution was mixed to achieve a mixed standard solution with a final concentration of 10 mg/L for each compound. Finally, extracts and standards were filtered through a 0.22 μm syringe filter and stored at −80 °C until UPLC-MS^E analysis.

3.4. Quantification of Total Phenolic Compounds

Total phenolic content was determined by the Folin–Ciocalteu method, which was adapted from Rocchetti et al. [52]. The extract and 2.5 mL Folin-Ciocalteu reagent solution 10% were combined and then mixed well using a Vortex. The mixture was allowed to react for 5 min then 2 mL of sodium carbonate 4% solution was added and mixed well the solution was incubated at room temperature (25 °C) in the dark for 2 h. The absorbance was measured at 750 nm using a spectrophotometer (Turner® 340, Haverhill, MA, USA) and the results were expressed in mg gallic acid equivalents (GAE) per 100 g of extract using a gallic acid (2,5–50 μg/μL) standard curve [53].

3.5. Phytochemical Profile Characterization by UPLC-MS^E

The UPLC-MS^E analyses were carried out according to Santos et al. (2018) [54] with slightly modifications. Two uL of extracts and standards were injected in triplicate into an Acquity UPLC (Waters, Milford, MA, USA) coupled to Xevo G2-S QTOF-MS/MS (Waters, Manchester, UK) system equipped with an electrospray ionization source (ESI) operating in negative ion mode. The column used was a UPLC HSS T3 C18 (100 mm × 2.1 mm, 1.8 μm) (Waters, Wexford, Ireland). The flow rate was 0.6 mL/min and the mobile phase gradient elution consisted of acidified water (5 mM ammonium formate and 0.3% formic acid, *v/v*) (pump A) and acetonitrile containing 0.3% formic acid (pump B), as follows: 97% A at 0 min, 50% A at 6.8 min, 15% A at 7.4–8.5 min, followed by an additional equilibration step 97% A at 9.1–12 min. Data were acquired using a multiplexed MS/MS acquisition in the extended mode with alternating low and high energy acquisition (MS^E) 30–55 eV using ultra-high pure argon (Ar). The capillary and cone voltage were set at 3.0 kV and 30 V, respectively. The desolvation gas (N$_2$) was set at 600 L/h at 450 °C, the cone gas was set at 50 L/h and the source temperature at 120 °C. Data acquisition was performed by using MassLynx 4.1 (Waters Corporation, Milford, MA, USA). To ensure accuracy and reproducibility, all acquisitions were performed by infusing lock mass calibration with leucine-enkephaline (Waters Corporation, Milford, MA, USA) (*m/z* 554.2615) at a concentration of 1.0 ng/L in acetonitrile: H$_2$O (50:50, *v/v*) with 0.1% (*v/v*) formic acid at a flow rate of 10 μL/min.

The raw data of all replicates were processed with Progenesis QI v2.1 (Nonlinear Dynamics, Waters Corporation, UK) with the following conditions: centroid data, resolution full-width at half maximum (FWHM) of 50,000, deprotonated molecule [M − H]$^-$. The identification of phenolic compounds was performed by searching for polyphenols with MetaScope, using a customized database of polyphenol compounds from PubChemID by using the following parameters: precursor and fragment mass error tolerance (5 and 10 mg/L, respectively), retention time limit 7.5 min. Target analysis was also applied for identification of the phenolic compounds by comparing the run parameters of phenolic standards such as the retention time, exact mass, mass error, and the MS-MS spectra, besides the other above mentioned parameters. The processed data were exported to XLSTAT software (Addinsoft, Paris, France), where the values of relative abundance obtained from each compound based on ion mass spectra counting were used to relative quantification and then to the multivariate statistical evaluation where Principal Components Analysis (PCA) biplot was elaborated. The Metaboanalyst 3.0 web server was used for analysis of multivariate data (HCA, hierarchical cluster analysis and heat map) where the eigenvectors of the correlation matrix was used to select the discriminatory bioactive compounds between the samples analyzed.

3.6. Analysis by Nuclear Magnetic Resonance Spectroscopy

For the NMR analyses, samples were prepared in triplicate. It was done by weighing 30.0 ($\pm$0.5) mg of freeze-dried samples (for each repetition), which was subjected during 10 min to shaking in a vortex with 0.8 mL of extractor solvent (D_2O for abrico and camapu; methanol-d4 for uxi), with TMSP-d4 at concentration 0.02% (*m/v*) as internal reference and mass standard and, after, to ultrasonic bath during 10 min. Next, samples were centrifuged (13,000 RPM) at room temperature for more 10 min. A 0.6 mL supernatant aliquot was analyzed.

The ^{1}H-NMR spectra, as well as the 2D experiments, were conducted in the Bruker Avance III 9.4 Tesla (400 MHz for hydrogen frequency) spectrometer equipped with a PABBI probe (5 mm) and with an Automatic Tuning and Matching (ATMA) unit located in NMR Laboratory of the Chemistry Department at Federal University of São Carlos-UFSCar, Brazil. The TMSP-d4 was used to determine the 0 point in the chemical shift scale. 1H NMR spectra were registered at temperature of 300 K using 64 K data points and the standard Bruker pulse sequence (zgps).

All spectra were obtained through pre-saturation pulse sequence in order to suppress the residual water sign in the solvent. Spectra were acquired through 16 scans (NS), spectral window 15.0191 mg/L with 128,000 data points, receiver gain (RG) 80.6, relation delay of 50 s (D1) and acquisition time of 10.9 s (AQ). The exponential line broadening function value adopted for spectral apodization (LB) was 0.30 Hz.

The NMR data was analyzed in TopSpin Brucker Software. Quantification was performed by comparison of the compound's signals area and the area of the TMSP-d4 signal to find out the compound's mass since the TMSP-d4 mass is known. The equation used was: analyte mass = (analyte area $\times$ TMSP-d4 ^{1}H N$^\circ$ $\times$ TMSP-d4 mass $\times$ analyte MM)/(TMSP-d4 area $\times$ analyte ^{1}H N$^\circ$ $\times$ TMSP-d4 MM). The quantification is expressed in mg/g of freeze-dried fruit.

3.7. ABTS

For ABTS assay, the procedure followed the method of Thaipong et al. [55]. The stock solutions included 7.0 mM ABTS solution and 140 mM potassium persulfate solution. The working solution was then prepared by mixing the two stock solutions, 5 mL ABTS solution in 88 µL potassium persulfate solution, and allowing them to react for 16 h at room temperature in the dark. The solution was then diluted by mixing 1 mL ABTS solution with ethanol necessary to obtain an absorbance of 0.709–0.701 units at 734 nm using the spectrophotometer (Turner® 340, Haverhill, MA, USA). The fruits extracts were allowed to react with 2.5 mL of the ABTS solution for 6 min in a dark condition. The standard curve was linear between 0 and 2000 µM Trolox equivalents. Results are expressed in µM Trolox equivalents g^{-1} of extract.

3.8. DPPH

The antioxidant activity of all fruits extracts in relation to the DPPH (2,2-diphenyl-1-picrilidrazil) radical was quantified by using a protocol described by Brand-Williams and Berset [56], using wave 515 nm in spectrophotometer (Turner® 340, Haverhill, MA, USA). The working solution was obtained by dissolving 2.4 mg DPPH reagent with 100 mL methanol and stored in the dark until needed. The extracts were allowed to react with 2.5 mL of the DPPH solution for 30 min in the dark. The standard curve was linear between 0 µM and 2000 µM Trolox. Results are expressed in µmol Trolox g^{-1} of extract.

3.9. FRAP

The FRAP assay was done according to Thaipong et al. [55]. The stock solutions included 0.3 M acetate buffer, pH 3.6, 10 mM TPTZ (2,4,6-tripyridyl-s-triazine) solution, and 20 mM ferric chloride solution. The fresh working solution was prepared by mixing 25 mL acetate buffer, 2.5 mL TPTZ solution, and 2.5 mL ferric chloride solution. Fruits extracts were allowed to react with 2.7 mL of the FRAP solution for 30 min in the dark condition and warmed at 37 °C. Readings of the colored product were then taken at 595 nm in spectrophotometer (Turner® 340, Haverhill, MA, USA). The standard curve was linear between 500 µM and 2000 µM ferrous sulphate. Results are expressed in µmol ferrous sulphate g^{-1} of extract.

3.10. ORAC

The ORAC procedure used an automated plate reader (SpectraMax i3x, Molecular Devices, USA) with 96-well plates [54–57]. Analyses were conducted in phosphate buffer pH 7.4 at 37 °C. Peroxyl radical was generated using 2, 2′-azobis (2-amidino-propane) dihydrochloride which was prepared fresh for each run. Fluorescein was used as the substrate. Fluorescence conditions were as follows: excitation at 485 nm and emission at 520 nm. The standard curve was linear between 1 µM and 90 µM Trolox. Results are expressed as µM TE g^{-1} of extract.

3.11. Statistical

Statistical comparisons were carried out by ANOVA and post hoc Tukey's test using Graph Pad Prism 5.0 and the differences were considered significant when $p < 0.05$ in total phenolic content and relative abundance of identified compounds. The correlation coefficients between different results of evaluation antioxidant capacity of the samples, ABTS, DPPH, FRAP, ORAC and total phenolic compounds (PC) were obtained. Principal component analysis (PCA) and heatmap was used to interpret data, using Stat graphics software.

4. Conclusions

This work highlights the significantly concentration of phenolic compounds and antioxidant properties of Amazonian fruits. A total of 293 phenolic compounds were tentatively identified in mamey apple, camapu, and uxi by comparison with standards and by fragmentation patterns. Mamey apple extracts presented major number of phenolic compounds, being the phenolic acids and terpenoids the classes more identified. ABTS, DPPH, FRAP, and ORAC assays revealed that the ethanol extract presents a higher antioxidant activity and aqueous extract presents a high correlation with phenolic compounds content in Amazonian fruits. The findings of this study highlight the potential of mamey apple, camapu, and uxi as a valuable source of natural antioxidants.

Supplementary Materials: The following are available online. Table S1: Bioactive compounds detected and characterized in mamey apple, camapu and uxi fruits extracted with aqueous (W) and ethanolic (E) solvents.

Author Contributions: Conceptualization, L.G.B.L., A.G.F., L.C.C., M.S.L.F., and A.J.T.; methodology, L.G.B.L., J.M., J.P.d.A., M.C.B.S., T.P.d.N., M.d.S.S., A.G.F, M.S.L.F., and A.J.T. data curation, L.G.B.L., J.M., J.P.d.A., M.C.B.S., T.P.d.N., M.d.S.S., A.G.F, M.S.L.F., and A.J.T. writing—original draft preparation, L.G.B.L., J.M., M.C.B.S., T.P.d.N., M.d.S.S., M.S.L.F., and A.J.T. writing—review and editing, L.G.B.L., M.C.B.S., T.P.d.N., M.S.L.F., and A.J.T. All authors have read and agreed to the published version of the manuscript.

Funding: This research was funded by FAPERJ grant number 248660, 246295, 216954 and 246296.

Acknowledgments: This work was supported by Coordenação de Aperfeiçoamento de Pessoal de Nível Superior (CAPES), Conselho Nacional de Desenvolvimento Científico e Tecnológico (CNPq), Fundação de Amparo à Pesquisa do Estado do Rio de Janeiro (FAPERJ) and Foundation for Research Support of the State of São Paulo (FAPESP).

Conflicts of Interest: The authors declare no conflict of interest.

Availability of Data and Materials: All the processed data are presented in the article. Information on raw data and materials are available from the corresponding authors upon request.

References

1. Ramos, A.S.; Souza, R.O.; Boleti, A.P.D.A.; Bruginski, E.R.; Lima, E.S.; Campos, F.R.; Machado, M.B. Chemical characterization and antioxidant capacity of the araçá-pera Psidium acutangulum: An exotic Amazon fruit. *Food Res. Int.* **2015**, *75*, 315–327. [CrossRef]

2. Cândido, T.L.N.; Silva, M.R.; Agostini-Costa, T.S. Bioactive compounds and antioxidant capacity of buriti (Mauritia flexuosa L.f.) from the Cerrado and Amazon biomes. *Food Chem.* **2015**, *177*, 313–319. [CrossRef] [PubMed]

3. Chisté, R.C.; Mercadante, A.Z. Identification and quantification, by HPLC-DAD-MS/MS, of carotenoids and phenolic compounds from the Amazonian fruit Caryocar villosum. *J. Agric. Food Chem.* **2012**, *60*, 5884–5892. [CrossRef] [PubMed]

4. Schulz, M.; Borges, G.D.S.C.; Gonzaga, L.V.; Seraglio, S.K.T.; Olivo, I.S.; Azevedo, M.S.; Nehring, P.; de Gois, J.S.; de Almeida, T.S.; Vitali, L.; et al. Chemical composition, bioactive compounds and antioxidant capacity of juçara fruit (Euterpe edulis Martius) during ripening. *Food Res. Int.* **2015**, *77*, 125–131. [CrossRef]

5. Bataglion, G.A.; da Silva, F.M.A.; Eberlin, M.N.; Koolen, H.H.F. Simultaneous quantification of phenolic compounds in buriti fruit (Mauritia flexuosa L.f.) by ultra-high performance liquid chromatography coupled to tandem mass spectrometry. *Food Res. Int.* **2014**, *66*, 396–400. [CrossRef]

6. Virgolin, L.B.; Seixas, F.R.F.; Janzantti, N.S. Composition, content of bioactive compounds, and antioxidant activity of fruit pulps from the Brazilian Amazon biome. *Pesqui. Agropecu. Bras.* **2017**, *52*, 933–941. [CrossRef]

7. Luís, E.; Ordóñez-Santos, G.M.; Martínez-Álvarez, A.M.V.-R. Effect of processing on the physicochemical and sensory properties of mammee apple (*Mammea americana* L.) fruit. *Agrociencia* **2014**, *48*, 377–385.

8. Sun, L.; Liu, J.; Liu, P.; Yu, Y.; Ma, L.; Hu, L. Immunosuppression effect of Withangulatin A from Physalis angulata via heme oxygenase 1-dependent pathways. *Process Biochem.* **2011**, *46*, 482–488. [CrossRef]

9. Hseu, Y.C.; Wu, C.R.; Chang, H.W.; Kumar, K.J.S.; Lin, M.K.; Chen, C.S.; Cho, H.J.; Huang, C.Y.; Huang, C.Y.; Lee, H.Z.; et al. Inhibitory effects of Physalis angulata on tumor metastasis and angiogenesis. *J. Ethnopharmacol.* **2011**, *135*, 762–771. [CrossRef]

10. Nunomura, R.C.S.; Oliveira, V.G.; Nunomura, S.M. Characterization of Bergenin in. *Amazonia* **2009**, *20*, 1060–1064.

11. Paz, W.H.P.; de Almeida, R.A.; Braga, N.A.; da Silva, F.M.A.; Acho, L.D.R.; Lima, E.S.; Boleti, A.P.A.; dos Santos, E.L.; Angolini, C.F.F.; Bataglion, G.A.; et al. Remela de cachorro (Clavija lancifolia Desf.) fruits from South Amazon: Phenolic composition, biological potential, and aroma analysis. *Food Res. Int.* **2018**, *109*, 112–119. [CrossRef] [PubMed]

12. de Brabo Sousa, S.H.; de Andrade Mattietto, R.; Campos Chisté, R.; Carvalho, A.V. Phenolic compounds are highly correlated to the antioxidant capacity of genotypes of Oenocarpus distichus Mart. fruits. *Food Res. Int.* **2018**, *108*, 405–412. [CrossRef] [PubMed]

13. Nonato, C.D.F.A.; Leite, D.O.D.; Pereira, R.C.; Boligon, A.A.; Ribeiro-filho, J.; Rodrigues, F.F.G.; Costa, J.G.M. Chemical analysis and evaluation of antioxidant and antimicrobial activities of fruit fractions of *Mauritia flexuosa* L.f. (Arecaceae). *PeerJ* **2018**, *6*, e5991. [PubMed]

14. Barros, R.G.C.; Andrade, J.K.S.; Denadai, M.; Nunes, M.L.; Narain, N. Evaluation of bioactive compounds potential and antioxidant activity in some Brazilian exotic fruit residues. *Food Res. Int.* **2017**, *102*, 84–92. [CrossRef] [PubMed]

15. Péroumal, A.; Adenet, S.; Rochefort, K.; Fahrasmane, L.; Aurore, G. Variability of traits and bioactive compounds in the fruit and pulp of six mamey apple (Mammea americana L.) accessions. *Food Chem.* **2017**, *234*, 269–275. [CrossRef] [PubMed]

16. Genovese, M.I.; Da Silva Pinto, M.; De Souza Schmidt Gonçalves, A.E.; Lajolo, F.M. Bioactive compounds and antioxidant capacity of exotic fruits and commercial frozen pulps from Brazil. *Food Sci. Technol. Int.* **2008**, *14*, 207–214. [CrossRef]

17. De Souza, V.R.; Pereira, P.A.P.; Queiroz, F.; Borges, S.V.; de Deus Souza Carneiro, J. Determination of bioactive compounds, antioxidant activity and chemical composition of Cerrado Brazilian fruits. *Food Chem.* **2012**, *134*, 381–386. [CrossRef]

18. Vasco, C.; Ruales, J.; Kamal-eldin, A. Total phenolic compounds and antioxidant capacities of major fruits from Ecuador pacific OCEAN. *Food Chem.* **2008**, *111*, 816–823. [CrossRef]

19. Rothwell, J.A.; Perez-jimenez, J.; Neveu, V.; Medina-remo, A.; Manach, C.; Knox, C.; Eisner, R.; Hiri, N.M.; Garci, P.; Wishart, D.S.; et al. Database Update Phenol-Explorer 3.0: A major update of the Phenol-Explorer database to incorporate data on the effects of food processing on polyphenol content. *Database* **2013**, *2013*, 1–8. [CrossRef]

20. Martins, N.; Barros, L.; Ferreira, I.C.F.R. In vivo antioxidant activity of phenolic compounds: Facts and gaps. *Trends Food Sci. Technol.* **2016**, *48*, 1–12. [CrossRef]

21. da Silva, C.P.; Freitas, R.A.M.S.; Sampaio, G.R.; Santos, M.C.B.; do Nascimento, T.P.; Cameron, L.C.; Simões, M.; Arêas, J.A.G. Identification and action of phenolic compounds of Jatobá-do-cerrado (Hymenaea stignocarpa Mart.) on α-amylase and α-glucosidase activities and flour effect on glycemic response and nutritional quality of breads. *Food Res. Int.* **2019**, *116*, 1076–1083. [CrossRef]

22. Cruz-Reys, A.; Chavarin, C.; Arias, M.P.C.; Taboada, J.; Jímenez, E. The Molluscacide activity of piquerol A isolated from Piqueria trinervia (Compositae) on 8 species of pulmonate snails. *Memórias Inst. Oswaldo Cruz* **1989**, *84*, 35–40. [CrossRef]

23. Medina, S.; Collado-González, J.; Ferreres, F.; Londoño-Londoño, J.; Jiménez-Cartagena, C.; Guy, A.; Durand, T.; Galano, J.M.; Gil-Izquierdo, A. Quantification of phytoprostanes–bioactive oxylipins–and phenolic compounds of Passiflora edulis Sims shell using UHPLC-QqQ-MS/MS and LC-IT-DAD-MS/MS. *Food Chem.* **2017**, *229*, 1–8. [CrossRef] [PubMed]

24. De Rosso, M.; Tonidandel, L.; Larcher, R.; Nicolini, G.; Dalla Vedova, A.; De Marchi, F.; Gardiman, M.; Giust, M.; Flamini, R. Identification of new flavonols in hybrid grapes by combined liquid chromatography-mass spectrometry approaches. *Food Chem.* **2014**, *163*, 244–251. [CrossRef] [PubMed]

25. Mezni, F.; Slama, A.; Ksouri, R.; Hamdaoui, G.; Khouja, M.L.; Khaldi, A. Phenolic profile and effect of growing area on Pistacia lentiscus seed oil. *Food Chem.* **2018**, *257*, 206–210. [CrossRef] [PubMed]

26. Black, C.A.; Parker, M.; Siebert, T.E.; Capone, D.L.; Francis, I.L. Terpenoids and their role in wine flavour: Recent advances. *Aust. J. Grape Wine Res.* **2015**, *21*, 582–600. [CrossRef]

27. Vithana, M.D.K.; Singh, Z.; Johnson, S.K. Levels of terpenoids, mangiferin and phenolic acids in the pulp and peel of ripe mango fruit influenced by pre-harvest spray application of FeSO4 (Fe2+), MgSO4 (Mg2+) and MnSO4 (Mn2+). *Food Chem.* **2018**, *256*, 71–76. [CrossRef]

28. Wu, S.; Dastmalchi, K.; Long, C.; Kennelly, E.J. Metabolite Pro fi ling of Jaboticaba (Myrciaria cauli fl ora) and Other Dark-Colored Fruit Juices. *J. Agric. Food Chem.* **2012**, *60*, 7513–7525. [CrossRef]

29. Choi, J.; Shin, K.M.; Park, H.J.; Jung, H.J.; Kim, H.J.; Lee, Y.S.; Rew, J.H.; Lee, K.T. Anti-inflammatory and antinociceptive effects of sinapyl alcohol and its glucoside syringin. *Planta Med.* **2004**, *70*, 1027–1032. [CrossRef]

30. Freitas, F.; Avio, A.; de Araújo, R.C.; Soares, E.R.; Nunomura, R.C.S.; Silva, F.M.A.; da Silva, S.R.S.; da Souza, A.Q.L.; de Souza, A.D.L.; de Franco-Montalbán, F.; et al. Biological evaluation and quantitative analysis of antioxidant compounds in pulps of the Amazonian fruits bacuri (Platonia insignis Mart), ing a (Inga edulis Mart), and uchi (Sacoglottis uchi Huber) by UHPLC-ESI-MS/MS Fl. *J. Food Biochem.* **2018**, *42*, e12455. [CrossRef]

31. Schiassi, M.C.E.V.; Souza, V.R.; de Lago, A.M.T.; Campos, L.G.; Queiroz, F. Fruits from the Brazilian Cerrado region: Physico-chemical characterization, bioactive compounds, antioxidant activities, and sensory evaluation. *Food Chem.* **2018**, *245*, 305–311. [CrossRef] [PubMed]

32. Moo-Huchin, V.M.; Moo-Huchin, M.I.; Estrada-León, R.J.; Cuevas-Glory, L.; Estrada-Mota, I.A.; Ortiz-Vázquez, E.; Betancur-Ancona, D.; Sauri-Duch, E. Antioxidant compounds, antioxidant activity and phenolic content in peel from three tropical fruits from Yucatan, Mexico. *Food Chem.* **2015**, *166*, 17–22. [CrossRef] [PubMed]

33. Oroian, M.; Escriche, I. Antioxidants: Characterization, natural sources, extraction and analysis. *Food Res. Int.* **2015**, *74*, 10–36. [CrossRef] [PubMed]

34. Dos Santos, M.; Mamede, R.; Rufino, M.; de Brito, E.; Alves, R. Amazonian Native Palm Fruits as Sources of Antioxidant Bioactive Compounds. *Antioxidants* **2015**, *4*, 591–602. [CrossRef]

35. Dávalos, A.; Carmem, G.-C.; Bartolomé, B. Extending Applicability of the Oxygen Radical Absorbance Capacity (ORAC—Fluorescein) Assay. *J. Agric. Food Chem.* **2004**, *52*, 48–54. [CrossRef]

36. Re, R.; Pellegreni, N.; Proeggente, A.; Pannala, A.; Yang, M.; Rice-Evans, C. Antioxidant Activity Applying an Improved Abts Radical Cation Decolorization Assay. *Free Radic. Biol. Med.* **1999**, *26*, 1231–1237. [CrossRef]

37. Kylli, P.; Nousiainen, P.; Biely, P.; Sipilä, J.; Tenkanen, M.; Heinonen, M. Antioxidant Potential of Hydroxycinnamic Acid Glycoside Esters. *J. Agric. Food Chem.* **2008**, *56*, 4797–4805. [CrossRef]

38. Pyrzynska, K.; Pekal, A. Application of free radical diphenylpicrylhydrazyl (DPPH) to estimate the antioxidant capacity of food samples. *Anal. Methods* **2013**, *5*, 4288–4295. [CrossRef]

39. Morales, J.C.; Lucas, R. *Structure—Activity Relationship of Phenolic Antioxidants and Olive Components*; Elsevier: Amsterdam, The Netherlands, 2010; ISBN 9780123744203.

40. Velika, B.; Kron, I. Original Article Antioxidant properties of benzoic acid derivatives against superoxide radical. *Free Radic. Antioxid.* **2012**, *2*, 62–67. [CrossRef]

41. Pasari, L.P.; Khurana, A.; Anchi, P.; Aslam Saifi, M.; Annaldas, S.; Godugu, C. Visnagin attenuates acute pancreatitis via Nrf2/NFκB pathway and abrogates associated multiple organ dysfunction. *Biomed. Pharmacother.* **2019**, *112*, 108629. [CrossRef]

42. Swanson, B.G. *Tannins and Polyphenols*; Elsevier: Amsterdam, The Netherlands, 2003; p. 5729.

43. Cidade, H.; Rocha, V.; Palmeira, A.; Marques, C.; Tiritan, M.; Ferreira, H.; Lobo, J.S.; Almeida, I.F.; Sousa, M.E.; Pinto, M. In silico and in vitro antioxidant and cytotoxicity evaluation of oxygenated xanthone derivatives. *Arab. J. Chem.* **2016**, *10*, 1–10. [CrossRef]

44. Dudonné, S.; Vitrac, X.; Coutière, P.; Woillez, M.; Mérillon, J.-M. Comparative Study of Antioxidant Properties and Total Phenolic Content of 30 Plant Extracts of Industrial Interest Using DPPH, ABTS, FRAP, SOD and ORAC Assays. *J. Agric. Food Chem.* **2009**, *57*, 1768–1774. [CrossRef]

45. Kader, A.A. Flavor quality of fruits and vegetables. *J. Sci. Food Agric.* **2008**, *88*, 1863–1868. [CrossRef]

46. Bochkov, D.V.; Sysolyatin, S.V.; Kalashnikov, A.I.; Surmacheva, I.A. Shikimic acid: Review of its analytical, isolation, and purification techniques from plant and microbial sources. *J. Chem. Biol.* **2012**, *5*, 5–17. [CrossRef]

47. Ross, A.C.; Caballero, B.; Cousins, R.J.; Tucker, K.L.; Ziegler, T.R. *Modern Nutrition in Health and Disease*, 11st ed.; Lippincott Williams & Wilkins: Philadelphia, PA, USA, 2014.

48. Akihiro, T.; Koike, S.; Tani, R.; Tominaga, T.; Watanabe, S.; Iijima, Y.; Aoki, K.; Shibata, D.; Ashihara, H.; Matsukura, C.; et al. Biochemical Mechanism on GABA Accumulation During Fruit Development in Tomato. *Plant Cell Physiol.* **2008**, *49*, 1378–1389. [CrossRef]

49. Solas, M.; Puerta, E.J.; Ramirez, M. Treatment Options in Alzheimer's Disease: The GABA Story. *Curr. Pharm. Des.* **2015**, *21*, 4960–4971. [CrossRef] [PubMed]

50. Nascimento, M.V.L. *Physalis Angulata Estimula Proliferação de Células-Tronco Neurais do Giro Denteado Hipocampal de Camundongos Adultos*; Universidade Federal do Pará: Belém, Brazil, 2013.

51. Giorgetti, M.; Negri, G. Plants from Solanaceae family with possible anxiolytic effect reported on 19th century's Brazilian medical journal. *Rev. Bras. Farmacogn.* **2011**, *21*, 772–780. [CrossRef]

52. Rocchetti, G.; Chiodelli, G.; Giuberti, G.; Masoero, F.; Trevisan, M.; Lucini, L. Evaluation of phenolic profile and antioxidant capacity in gluten-free flours. *Food Chem.* **2017**, *228*, 367–373. [CrossRef] [PubMed]

53. Saeed, N.; Khan, M.R.; Shabbir, M. Antioxidant activity, total phenolic and total flavonoid contents of whole plant extracts *Torilis leptophylla* L. *BMC Complement Altern. Med.* **2012**, *12*, 221. [CrossRef]

54. Santos, B.M.C.; da Silva Lima, L.R.; Nascimento, F.R.; do Nascimento, T.P.; Cameron, L.C.; Ferreira, M.S.L. Metabolomic approach for characterization of phenolic compounds in different wheat genotypes during grain development. *Food Res. Int.* **2019**, *124*, 118–128. [CrossRef]

55. Thaipong, K.; Boonprakob, U.; Crosby, K.; Cisneros-Zevallos, L.; Hawkins Byrne, D. Comparison of ABTS, DPPH, FRAP, and ORAC assays for estimating antioxidant activity from guava fruit extracts. *J. Food Compos. Anal.* **2006**, *19*, 669–675. [CrossRef]

56. Brand-Williams, W.; Cuvelier, M.E.; Berset, C. Use of a free radical method to evaluate antioxidant activity. *LWT Food Sci. Technol.* **1995**, *28*, 25–30. [CrossRef]

57. Prior, R.L.; Hoang, H.; Gu, L.; Wu, X.; Bacchiocca, M.; Howard, L.; Hampsch-Woodill, M.; Huang, D.; Ou, B.; Jacob, R. Assays for Hydrophilic and Lipophilic Antioxidant Capacity (oxygen radical absorbance capacity (ORAC FL)) of Plasma and Other Biological and Food Samples. *J. Agric. Food Chem.* **2003**, *51*, 3273–3279. [CrossRef] [PubMed]

Sample Availability: Samples of the compounds are not available from the authors.

Article

Optimization of Scorpion Protein Extraction and Characterization of the Proteins' Functional Properties

Ahmidin Wali [1,2], Atikan Wubulikasimu [1,2], Sharafitdin Mirzaakhmedov [3], Yanhua Gao [1], Adil Omar [1,2], Amina Arken [1,2], Abulimiti Yili [1,*] and Haji Akber Aisa [1]

[1]   Key Laboratory of Plant Resources and Chemistry in Arid Regions, Xinjiang Technical, Institute of Physics and Chemistry, Chinese Academy of Sciences, Urumqi 830011, Xinjiang, China; ahmidin@ms.xjb.ac.cn (A.W.); obuatikam@sina.com (A.W.); gaoyh@ms.xjb.ac.cn (Y.G.); adilomar@126.com (A.O.); ak950208@163.com (A.A.); haji@ms.xjb.ac.cn (H.A.A.)
[2]   University of Chinese Academy of Science, Beijing 100039, China
[3]   Institute of Bioorganic Chemistry, Academy of Sciences of Uzbekistan, Tashkent 100125, Uzbekistan; mirzaakhmedov@mail.ru
*   Correspondence: abu@ms.xjb.ac.cn; Tel.: +86-991-3835679

Academic Editor: Jose M. Miranda
Received: 26 September 2019; Accepted: 6 November 2019; Published: 13 November 2019

Abstract: Scorpion has long been used in traditional Chinese medicine, because whole scorpion body extract has anti-cancer, analgesic, anti-thrombotic blood anti-coagulation, immune modulating, anti-epileptic, and other functions. The purpose of this study was to find an efficient extraction method and investigate some of physical and chemical parameters, like water solubility, emulsification, foaming properties, and oil-holding capacity of obtained scorpion proteins. Response surface methodology (RSM) was used for the determination of optimal parameters of ultrasonic extraction (UE). Based on single factor experiments, three factors (ultrasonic power (w), liquid/solid (mL/g) ratio, and extraction time (min)) were used for the determination of scorpion proteins (SPs). The order of the effects of the three factors on the protein content and yield were ultrasonic power > extraction time > liquid/solid ratio, and the optimum conditions of extraction proteins were as follows: extraction time = 50.00 min, ultrasonic power = 400.00 w, and liquid/solid ratio = 18.00 mL/g. For the optimal conditions, the protein content of the ultrasonic extraction and yield were 78.94% and 24.80%, respectively. The solubility, emulsification and foaming properties, and water and oil holding capacity of scorpion proteins were investigated. The results of this study suggest that scorpion proteins can be considered as an important ingredient and raw material for the creation of water-soluble supramolecular complexes for drugs.

Keywords: scorpion (*Buthus martensii* Karsch) protein; ultrasonic extraction; response surface methodology; scanning electron microscopy; functional properties

---

1. Introduction

Scorpions are Chelicerate arthropods and members of the class Arachnida. There are about 1500 known scorpion species, and only 30 among them produce venom considered potentially dangerous to humans, which might be lethal without medical treatment [1,2]. Scorpion venoms are rich sources of complex mixtures of substances, including toxic peptides, free amino acids, enzymes, nucleotides, lipids, amines, mucoproteins, heterocyclic components, and inorganic salts, which affect the ion channels of both excitable and non-excitable cells [3,4]; some scorpion venoms also contains bioactive compounds, like trimethylamine, nucleosides, and betaine. Therefore, it has anti-thrombosis, anticoagulant, fibrinolysis, analgesic, anti-tumor, anti-epileptic, and several pharmacological effects [5–8]. A number of studies indicate that scorpion venom is comprised of more than 300 toxins, most of which are less than 10 kDa in molecular mass [9]. Scorpions have been used in traditional medicine in Asia and Africa for thousands of years, and large-scale commercial farming has achieved good social and economic benefits [10,11]. The body parts of scorpions are effective for the treatment of cancer [12–14]. In this work, our main attention is focused on investigating whole scorpion body extract, not directed to venomous part.

With the emergence of several novel methods for the study and identification of scorpion body parts and venom components, several bioactive peptides have been proved effective to treat a variety of diseases [15–17]. The whole body part of a scorpion is used as medicinal material in traditional Chinese medicine. Although there are previous studies on the extraction, isolation, and activity of scorpion toxin [18,19], few reports are available on the extraction and functional evaluation of scorpion total protein. Scorpion bodies are a main medicinal recourse, but few studies on the bioactive peptides from scorpion proteins have been mentioned [20].

Most of the functional characteristics of proteins play a significant role in the protein's physical and chemical properties in the processing, storage, preparation, and sale phase at different areas of economy [21,22]. Many factors, such as the protein's size, pH, temperature, protein's shape, ionic strength, structure, amino acid sequence, composition, and charge distribution are very important factors of protein functionality. The conditions of the proteins at extraction and each stage of purification and drying are factors that must be taken into consideration [23,24]. Therefore, the aim of this study was to investigate some of physical and chemical parameters, like water solubility, emulsification, foaming properties, and oil-holding capacity of scorpion proteins. In order to reach of this goal, response surface methodology for the optimization of three parameters of ultrasonic extraction (UE)—ultrasonic power, solid/liquid ratio, and extraction time—are also investigated.

2. Materials and Methods

2.1. Materials and Chemicals

Scorpions (*Buthus martensii* Karsch) were collected (about 1000 pieces) in April in the Turpan region, Xinjiang, China, and after killing, they were put into plastic bottles and kept in a refrigerator. A Pierce BCA Protein Assay Kit was purchased from Thermo Scientific; electrophoresis reagents ware purchased from Biosharp Corporation (Beijing, China). All other chemicals were purchased from local suppliers.

The freeze drier was an FDU-1110 (EYELA Company, Tokyo, Japan), the high-speed refrigerated centrifuge was a CR22N (Hitachi Koki Co., Ltd., Tokyo, Japan), the refrigerated centrifuge was an 5417R (Eppendorf, Germany), and the large-capacity oscillator was an HY-8A (Millipore, Burlington, MA, USA). A Spectra Max M5 enzyme labeling analyzer (Bio-Tec Co., Ltd., USA) was also used.

2.2. Total Extraction of Proteins from Scorpion Bodies

Ten grams of dried scorpion were ground and separated through a 40-mesh separator. After that, separated powders (8.2 g) were subjected to de-oiling by petroleum ether (150 mL) in a continuous soxhlet extractor, until the solvent became colorless. The final residue was collected and air-dried. According to the following procedure, scorpion proteins (SPs) were extracted by (1) distilled water, (2) sodium chloride (0.5 M), (3) phosphate buffer (PBS; 50 mM, pH = 8.0), and (4) isoelectric precipitation (adjusted pH to 9.0 with 1.0 N NaOH, and adjusted pH to 4.5 with 1.0 N HCl to precipitate SPs) using an ultrasonic extraction method and a stirring extraction method (as a comparison) for 1 h, respectively. This method was used for the first time by us. Scorpion proteins were extracted with an ultrasonic generator, and the ultrasonic power was 200 w. The extraction steps were repeated three times and dialyzed against distilled water for 48 h at 4 °C (F0136-1 Dialysis Membrane, MWCO 1000Da, United States). The obtained dialysates were lyophilized using a freeze drier (FDU-1110, EYELA Company, Tokyo, Japan) and kept at −20 °C until use.

2.3. Methods for the Determination of Protein Extraction Indexes

2.3.1. Determination of Protein Contents (%)

The protein content of lyophilized SPs was determined by the BCA (bicinchoninic acid) method [25]. Bovine serum albumin (BSA) was used as a standard. A total of 1.0 mL of distilled water was used to dissolve 1.0 g of sample, which was clarified through centrifugation at 4500× g for 15 min at 4 °C. According to the specification of the BCA protein measuring kit, the absorbance was read by an enzyme labeling instrument (Spectra Max M5 enzyme labeling analyzer) at 562 nm. The average value of three parallel measurements was used to calculate the protein content.

2.3.2. Determination of Yields (%)

The yield of SPs was determined using Equation (1):

$$Yield\ (\%) = \frac{m_{Se}}{m_{Sd}} \times 100\% \tag{1}$$

where m_{Se} is the weight of scorpion protein extraction (mg), and m_{Sd} is the weight of the defatted scorpion (mg).

2.4. Single-Factor Experiments

The influence of ultrasonic power (w), extraction time (min) and liquid/solid ratio (mL/g) on protein content and yield of the UE were investigated. Based on the ultrasonic extraction and stirring extraction experiments, scorpion protein was extracted with 0.5 M NaCl buffer by the ultrasonic extraction method. The ultrasonic extraction time (20, 30, 40, 50, and 60 min), liquid/solid ratio (5, 10, 15, 20, and 25 mL/g), and ultrasonic power (50, 100, 200, 300, and 400 w) were tested.

2.5. Gel Electrophoresis Analysis

Gel electrophoresis (SDS-PAGE) of the SPs was done using 15% acrylamide gel (Bio-Rad Mini-PROREAN Tetra System) [26]. Gel electrophoresis was carried out at 10 v/cm constant voltage. Coomassie Brilliant Blue R-250 dye was used for staining the protein bands for 1.5 h, which were then de-stained in a decoloring solution for 1.5 h via shaking by large capacity oscillator (HY-8A).

2.6. Box–Behnken of RSM and Statistical Analysis

RSM is a statistical tool for solving multivariable problems by using reasonable experimental design method and obtaining certain data through experiments, and by analyzing regression equations to determine the optimal technological parameters [27]. In this study, the RSM and Box–Behnken design was used to optimize the extraction conditions. Based on the single-factor experiment, the complete design was made up of 17 runs, and these were done in triplicate to optimize the levels of the selected variables (extraction time, ultrasonic power, and liquid/solid ratio). For the statistical analysis, the three independent variables were coded as X_1, X_2, and X_3, respectively using Equation (2):

$$x_i = \frac{X_i - X_0}{\Delta X_i} \tag{2}$$

where x_i is the coded value of an independent variable, X_i is the real value of an independent variable, X_0 is the real value of an independent variable at the center point, and ΔX_i is the step change value.

The quadratic equation of the variables is as follows:

$$Y = \beta_0 + \sum \beta_i x_i + \sum \beta_{ii} x_i^2 + \sum \beta_{ij} x_i x_j \tag{3}$$

where Y is the predicted response variable; B_0, β_i, β_{ii}, and β_{ij} are the constant regression coefficients of the model; and x_i and x_j ($i = 1, 3; j = 1, 3, i \neq j$) represent the independent variables.

The accuracy and fitness of the above model were evaluated by the coefficient of determination (R^2) and the F-value. Based on the above results, the second order polynomial coefficients were undertaken using Design Expert (Version 8.0.6, USA) software. The model was performed to evaluate the analysis of variance (ANOVA).

2.7. Functional Properties

The functional properties of de-oiled scorpion flour (DSF), stirring extraction (SE), and ultrasonic extraction (UE) have been tested at the optimal conditions.

2.7.1. Protein Solubility Analyses

Using the method from [28], with slight modification, the solubility of the protein was determined. A total of 200 mg of SP was dissolved in 20 mL deionized water, and the pH was adjusted to 2, 3, 4, 5, 6, 7, 8, 9, 10, 11, and 12 by 1.0 mol/L HCl and 1.0 mol/L NaOH. After the pH was stabilized, it was stirred for 0.5 h at a speed of 8000 rpm/min for 20 min; centrifugation was done to obtain the supernatant. Using the Bradford method [29], the protein content of the supernatant was determined. The value of each sample was measured three times, and the average value was taken. Protein solubility was calculated using the following formula:

$$Protein \text{ Solubility } \% = \left(\frac{m}{m_T}\right) \times 100\% \tag{4}$$

where m is the protein content of the supernatant (mg/g), and m_T is the content of total protein in SP (mg/g).

2.7.2. Foaming Properties

The foaming capacity (FC) and foaming stability (FS) were carried out using the methods described previously, with minor modifications [24,30]. Using these methods, 0.2 g of SP was dissolved in 20mL of distilled water ($w/v = 1.0\%$), and the solution (V_0) was stirred for 2.0 min. The mixture was transferred immediately into a 50 mL tube, the initial foam volume was measured (V_1), and the foam volume after standing for 30 min was also measured (V_3). The FC was calculated using Equation (5):

$$FC\ (\%) = \left(\frac{V_1 + V_2 - V_0}{V_0}\right) \times 100\% \tag{5}$$

where V_2 is the volume of the liquid remaining just after stirring.

FS is calculated using the following formula:

$$FS\ (\%) = \frac{V_3}{V_1} \times 100\%. \tag{6}$$

2.7.3. Water Absorption Capacity (WAC)

Using the method from [31], with slight modifications, water absorption capacity (WAC) was determined. Here, 1 g (W_0) of SP was weighed together with the centrifugal tube (W_1). Thereafter, 10 mL of distilled water was added in the tube and mixed. After being held at room temperature for 30 min, the emulsion was centrifuged at 2100× g for 15 min. In the end, the supernatant was decanted, and the tube with sediment was weighed (W_2). Using Equation (7), the WAC was calculated:

$$WAC = \frac{W_2 - W_1}{W_0}. \tag{7}$$

2.7.4. Oil Absorption Capacity (OAC)

The oil absorption capacity (OAC) was determined using the method reported previously [31], with minor modifications. One gram (F_0) of SP was weighed together with a centrifugal tube (F_1) and mixed with 5 mL of corn oil. The emulsion was incubated at room temperature for 30 min, and then centrifuged (Eppendorf centrifuge 4530R, Germany) at 3500 rpm/min for 15 min under the same conditions. The tube was reweighed (F_2) after the removal of the supernatant. The OAC was determined using Equation (8):

$$\mathrm{OAC} = \frac{F_2 - F_1}{F_0} \tag{8}$$

2.7.5. Emulsifying Properties

The emulsifying activity (EA) of the UE and SE was determined by the method described by Wu et al. [32], with slight modification. Here, 1.0 g of each SP was weighed and dissolved in 20 mL of distilled water and 20 mL of corn oil, and mixed at a high speed for 1 min at room temperature. The mixture was centrifuged at 4000 rpm for 5 min. The EA was calculated using Equation (9):

$$\mathrm{EA}\ (\%) = \frac{h}{H} \times 100\% \tag{9}$$

where h is the height of the emulsified layer and H is the height of the tube contents.

The emulsion stability (ES) of UE and SE were determined using the method from [30], with slight modifications. One gram of each SP was weighed and dissolved with 20 mL of distilled water and 20 mL of corn oil, and thereafter mixed at a high speed for 2 min at room temperature. The mixture was centrifuged at 4000 rpm for 5 min. After preparation, 30 mL of emulsions was then transferred into test tubes, and the emulsions were stored at room temperature; they separated into a top oil layer and a bottom serum layer over time. The ES was evaluated using Equation (10):

$$\mathrm{ES}\ (\%) = \frac{H_s}{H_t} \times 100\% \tag{10}$$

where H_s is the height of bottom serum layer (mm) and H_t is the total height of emulsion in the tube (mm).

2.8. Scanning Electron Microscopy (SEM) Analysis

To study the effect of scorpion protein extraction technology on protein content and yield, we analyzed SPs by scanning electron microscopy (SEM; SUPRA 55VP, ZEISS). The morphology and surface characteristics of the SPs were observed and recorded by SEM after the samples were fixed on silicon wafers and sputtered by gold.

2.9. Statistical Analysis

The experiments were performed with three independent trials, and all the determinations were triplicated. The results were represented as mean ± standard deviation. Analysis of variance (ANOVA) was performed to identify significant differences ($p < 0.05$).

3. Results and Discussion

3.1. Analysis of Scorpion Protein Extraction Method

In this study, the effects of diverse buffer solutions for the extraction of scorpion protein were studied. Meanwhile, the effects of ultrasonic extraction (this method is used to increase the yield of protein), as well as stirring extraction of the yield (%) and protein content (%) were compared. Figure 1 and Table 1 show that the order of the effects of four buffer solutions on yield and protein content was 0.5 M NaCl > 20 mM PBS > 0.02 M NaOH > water. The 0.5 M NaCl buffer solution (yield 14.64 ± 0.08%, protein content 79.06 ± 0.05%) with ultrasonic was better than other buffers for extracting scorpion protein, followed by 20 mM PBS (yield 18.29 ± 0.05%, protein content 60.98 ± 0.07%).

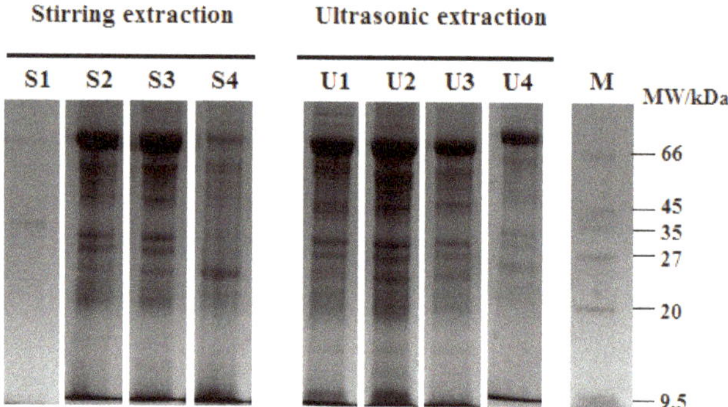

Figure 1. SDS-PAGE electrophoresis of scorpion proteins with different extractions and buffer conditions. M refers to the marker; S1, S2, S3, and S4 are scorpion proteins extracted by stirring with water, 0.5 M NaCl, 20 mM phosphate buffer (PBS), alkali extraction, and isoelectric precipitation, respectively; U1, U2, U3, and U4 are scorpion proteins extracted by ultrasonic with water, 0.5 M NaCl, 20 mM PBS, alkali extraction, and isoelectric precipitation, respectively.

Table 1. Effects of ultrasonic and stirring methods on extraction of total proteins from scorpion body.

Extraction Method	Yield (%)		Protein Content (%)	
	Ultrasonic Extraction	Stirring Extraction	Ultrasonic Extraction	Stirring Extraction
Water	34.85 ± 0.06	34.15 ± 0.09	31.14 ± 0.04	18.08 ± 0.06
0.5 M NaCl	14.64 ± 0.08	11.80 ± 0.03	79.06 ± 0.05	35.26 ± 0.08
20 mM PBS	18.29 ± 0.05	13.45 ± 0.10	60.98 ± 0.07	37.25 ± 0.09
0.02 M NaOH	7.70 ± 0.11	7.57 ± 0.13	50.87 ± 0.07	51.91± 0.17

3.2. Analysis of a Single Factor Results

3.2.1. Effect of Extraction Time on Yield and Protein Content

Extraction time plays a key role in protein extraction [32]. In this experiment, the effect of extraction time on the yield and protein content is examined. For that, ultrasound power was 200 W, the liquid/solid ratio was 15, and extraction time was carried out in a range between 20–60 min. The results obtained are shown in Figure 2A. Obtained data show that with the increase of ultrasonic extraction time, the yield and protein content rapidly increased and reached a maximum at 50 min, at 17.95% and 66.35%, respectively. A further increase of time did not affect the yield and amount of protein.

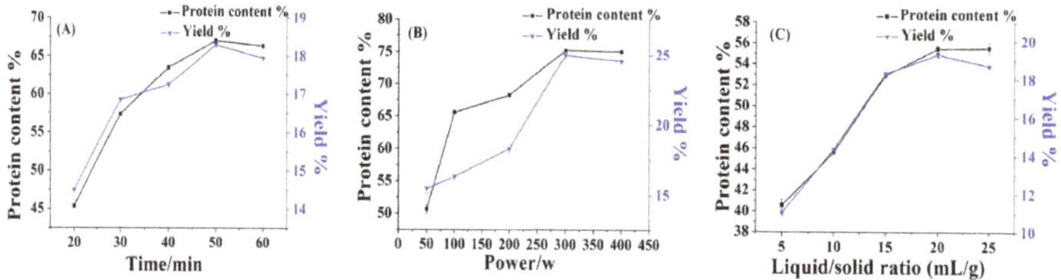

Figure 2. Effects of extraction time (**A**), ultrasonic power (**B**), and liquid/solid ratio (**C**) on yield and protein content.

3.2.2. Effect of Ultrasonic Power on Yield and Protein Content

In this experiment, the effect of ultrasonic power on the yield and protein content was examined. For that, the extraction time was 50 min, the liquid/solid ratio (mL/g) was 15, and the ultrasonic power was evaluated in the range 50–400 w. The results obtained are shown in Figure 2B. The obtained data show that with an increase in power, the yield and protein content increased rapidly, and reached their maximums at 300 w, at 23.60% and 75.30%, respectively; in addition, a further increase of time did not affect the yield and amount of protein.

3.2.3. Effect of Liquid/Solid Ratio on Yield and Protein Content

In this experiment, the effect of the liquid/solid ratio on the yield and protein content is examined. For that, ultrasound power was 200 w, extraction time was 50 min, and the liquid/solid ratio was evaluated in the range between 5–25 mL/g. The results obtained are shown in Figure 2C. The obtained data show that with the increase of liquid/solid ratio to the 15 mL/g the yield, protein content rapidly increased, and when the ratio reached to 20 mL/g, a slower increase can be seen, reaching maximums of 19.21% and 55.32%, respectively. A further increase of time did not affect the yield and amount of protein.

3.3. Optimization of Extraction Parameters by RSM

Table 2 shows that there was a significant change in protein content and yield at different values of the selected parameters. Using Equation (3), the results were analyzed. After multiple regressions fitting with Design-Expert V8.0.6 software, the regression model equation is as follows:

$$Y_1 = 56.43 + 7.25A + 0.066B + 7.69C + 4.80AB + 4.97AC + 4.67BC + 2.14A^2 + 1.28B^2 - 3.18C^2 \quad (11)$$

$$Y_2 = 19.46 + 0.86A + 0.75B + 1.93C + 1.14AB + 1.80AC + 1.72BC + 0.31A^2 - 1.64B^2 - 0.26C^2 \quad (12)$$

where Y_1 and Y_2 are the predicted conversion (%) of protein content and yield, respectively; and A, B, and C are the extraction time (X_1, min), liquid/solid ratio (X_2, mL/g), and ultrasonic power (X_3, w), respectively.

When the factor sign is positive, it shows that the amount of the response variable increases with an increase in its value, and vice versa. Figure 3A,C can also be used to compare the model predicted and the experimental results. Figure 3B,D is the normal probability chart indicating that the points follow a narrow linear pattern. The analysis of variance of the regression equation is shown in Table 3. F is the value of the regression model, where the protein content and yield were 17.49 and 34.12 and the p-values were 0.0005 and <0.0001, respectively. These values show that the model obtained was significant. The F-value and p-value for the lack of fit of the protein content and yield were 1.54 and 2.04, respectively, and 0.3354 and 0.2506, respectively, which shows that the equation has a good fitting degree and high reliability. In addition, the decision coefficients (R^2) of the model were 0.9574 and 0.9777, respectively, and the adjusted coefficients of determination (Adj-R^2) were 0.9027 and 0.9491, respectively. The values of R^2 and adjusted R^2 for the models are shown in Table 3, which shows that the regression equation can predict the result of extracting scorpion protein accurately.

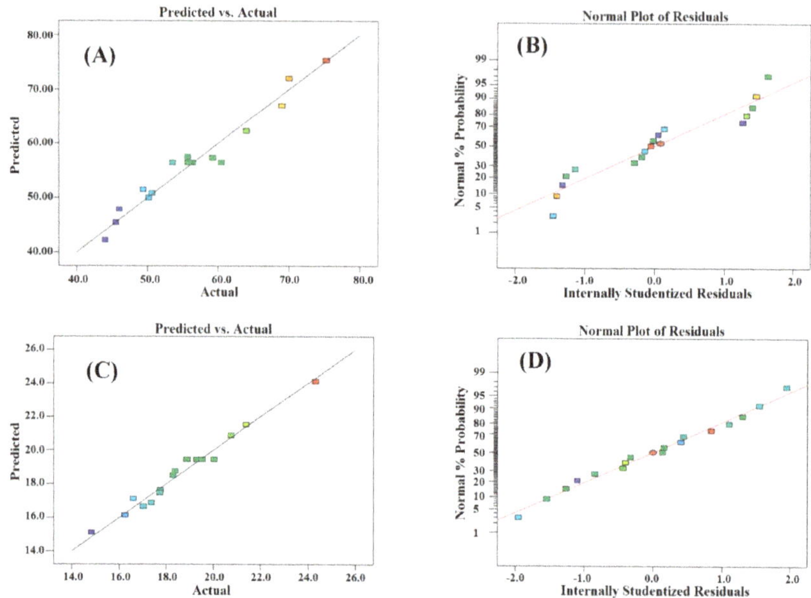

Figure 3. Standard statistical diagrams for model verification. (**A,C**) Model-predicted values versus actual data. This figure shows a comparison between the model-predicted values and the actual values that have been obtained by the experiments. (**B,D**) Normal probability plot of the residuals. Monotonous distribution and linearity of the data, which are obvious from these figures, confirm the validity of the model and its capability to predict the results.

Table 2. RSM experimental design and results for the three-factor/three-level Box–Behnken design (BBD).

Runs	Extraction Time (min) X_1	Liquid/Solid Ratio (mL/g) X_2	Ultrasonic Power (W) X_3	Protein Content/%	Yield /%
1	60	15	400	75.23	24.32
2	40	20	400	69.01	21.41
3	20	10	300	59.31	17.76
4	40	15	300	55.98	19.53
5	40	15	300	56.39	18.89
6	60	10	300	64.09	16.62
7	40	20	200	44.01	14.84
8	40	10	200	49.4	16.25
9	60	15	200	50.17	17.04
10	40	10	400	55.70	17.74
11	40	15	300	55.70	19.31
12	60	20	300	70.00	20.78
13	40	15	300	53.58	19.52
14	20	20	300	46.01	17.37
15	20	15	200	45.49	18.31
16	20	15	400	50.66	18.39
17	40	15	300	60.52	20.04

Table 3. Least-squares fit and parameter estimates (significance of regression coefficient).

Source	Protein Content		Yield	
	F-Value	*p*-Value	*F*-Value	*p*-Value
Model	17.49	0.0005	34.12	< 0.0001
X_1	53.50	0.0002	23.98	0.0018
X_2	4.46	0.9486	18.15	0.0037
X_3	60.17	0.0001	118.71	< 0.0001
$X_{1 \times 2}$	11.73	0.0111	20.67	0.0026
$X_{1 \times 3}$	12.57	0.0094	51.76	0.0002
$X_{2 \times 3}$	11.11	0.0125	25.77	0.0014
X_1^2	2.45	0.1617	1.67	0.2378
X_2^2	0.88	0.3800	45.25	0.0003
X_3^2	5.43	0.0526	1.12	0.3256
Lack of Fit	1.54	0.3354	2.04	0.2506
R^2	0.9574		0.9777	
Adj-R^2	0.9027		0.9491	

3.3.1. Analysis of the Influence of Various Factors on the Extraction of Scorpion Proteins

The degree of influence of each factor on the test index can be compared by the *F*-value (Table 4): $F (X_1) = 53.50, 23.98$; $F (X_2) = 4.46, 18.15$; $F (X_3) = 60.17, 118.71$. These values indicate that the order of influence of each factor on the extraction of scorpion protein was ultrasonic power (w) > extraction time (min) > liquid/solid ratio (mL/g). The response surface can describe the interaction between variables and predict the optimal conditional values of each factor [33]. Figure 4A–F shows the effect of operational variables on protein content and yield protein content and yield. The effect of extraction time on the protein process is greater than that of the ratio of the liquid/solid, and that of ultrasonic power is greater than that of extraction time. According to the above analysis, the order of the influence of each variable on the extraction technology of scorpion protein was ultrasonic power > extraction time > liquid/solid ratio.

3.3.2. Interactions of Variables

Contour plots for each of the fitted models that display the effects of the three variables (to visualize the combined effects of the three factors on protein content and yield) were generated. Figure 4 illustrates the two-dimensional (2D) plots of the binary interactions of the variables on protein content and yield via the contour plots. The interaction between the extraction time and the liquid/solid ratio is shown in Figure 4A,D. This plot indicates that protein content and yield depend more on X_1 than on X_2. Figure 4A,D reveals that at low values of X_1, maximum protein content and yield occurs at higher values of X_2. However, at higher values of X_1, maximum protein content and yield occurs at lower values of X_2. As seen in Figure 4A,D, the interaction between the two factors is weak. The interaction between the extraction time and ultrasonic power is shown in Figure 4B,E. This plot indicates that protein content and yield depend more on X_3 than on X_1. In Figure 4C,F, the effect of the interaction between the liquid/solid ratio and ultrasonic power is depicted. Figure 4C,F shows that increasing the liquid/solid ratio at different ultrasonic power levels has no important effect on the protein content and yield, so the plot shows that protein content and yield depends more on X_3 than on X_2. Therefore, the optimum values of X_1, X_2, and X_3, as determined by the software, are 50 min, 400 w, and 18 mL/g, respectively.

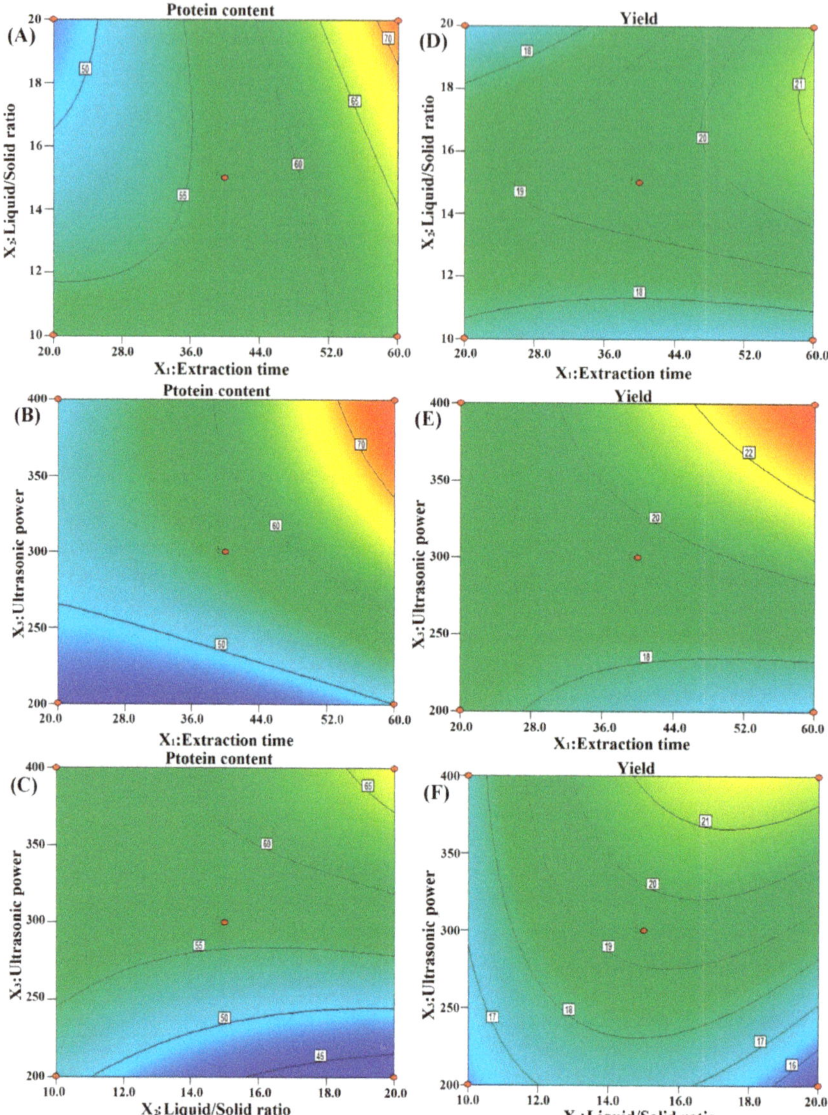

Figure 4. Effects of operational variables on protein content and yield. Interaction between (**A**) liquid/ solid ratio and extraction time, (**B**) ultrasonic power and extraction time, and (**C**) ultrasonic power and liquid/solid ratio to protein content. Also depicted are the interaction between the (**D**) liquid/solid ratio and extraction time, (**E**) ultrasonic power and extraction time, and (**F**) ultrasonic power and liquid/solid ratio to yield.

3.3.3. Determination and Validation of Optimal Extraction Conditions

Optimum conditions, which were obtained by Design-Expert V8.0.6 software, were as follows: extraction time = 47.68 min; ultrasonic power = 395.84 w; and solid/liquid ratios = 18.01 mL/g. In view of the feasibility of the experiment, the optimum conditions of extraction proteins were adjusted as follows: extraction time = 50.00 min, ultrasonic power = 400 w, and solid/liquid ratio = 18.00 mL/g.

After several tests ($n > 3$), the protein content and yield were 78.94%, and 24.80%, respectively. It is shown that the regression equation and the optimal conditions obtained by the response surface method are reliable.

3.4. Functional Properties of Scorpion Proteins

3.4.1. Protein Solubility (PS)

The protein solubility (PS) profiles of de-oiled scorpion flour (DSF), ultrasonic extraction (UE), and stirring extraction (SE) in the water at different pH ranges (2–12) are shown in Figure 5. The PS of DSF, UE, and SE were significantly different, and Figure 5 shows the same U-shaped curves. When the pH value is in the range of 2–4, the solubility of DSF, UE, and SE decreased; however, when in the pH range of 6–10, the solubility of DSF, UE, and SE significantly increased. The minimum protein solubility of DSF, UE, and SE were presented at pH 4, with values of 8.05%, 15.25%, and 18.75%, respectively. The maximum protein solubility was presented at pH 12, with values of 13.5%, 70.15%, and 79.5%, respectively. Therefore, results indicate that scorpion protein extracted by ultrasonic method shows exceptional solubility at an alkaline pH.

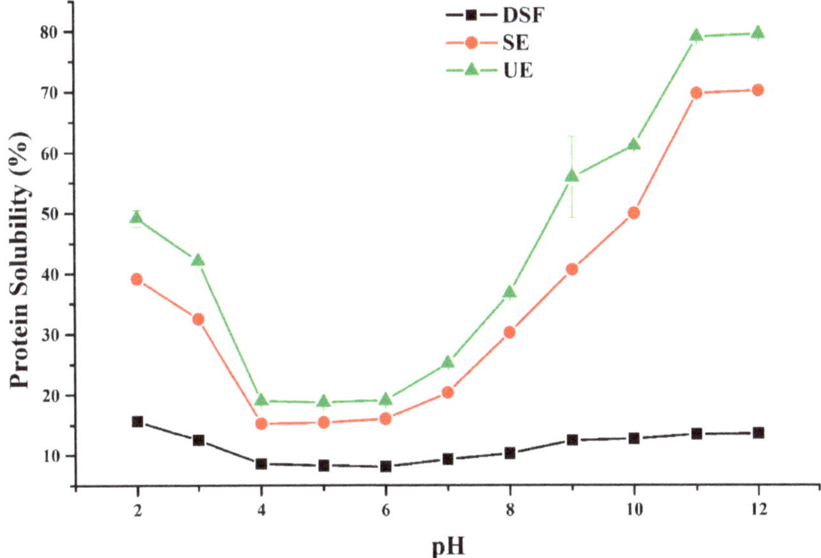

Figure 5. Protein solubility profiles of de-oiled scorpion flour (DSF), stirring extraction (SE), and ultrasonic extraction (UE) at different pH levels.

3.4.2. Water and Oil Absorption Capacity

The obtained results for the water absorption capacity (WAC) and oil absorption capacity (OAC) of proteins extracted by ultrasonic and stirring are presented in Table 4. The WAC and OAC from both methods were significantly different ($p < 0.05$), with UE having the higher WAC (40.3 ± 0.28 g/g) and OAC (27.70 ± 0.14 g/g) than SE (with WAC = 33.45 ± 0.07 g/g and OAC = 18.80 ± 0.15 g/g) at pH 7.0.

Table 4. Functional properties of scorpion proteins.

Properties	Ultrasonic Extraction	Stirring Extraction
Water holding capacity (g/g)	25.25 ± 0.21	20.45 ± 0.07
Oil holding capacity (g/g)	38.50 ± 0.14	30.65 ± 0.64
Emulsifying activity (%)	45.55 ± 0.64	40.25 ± 0.07
Emulsion stability (%)	85.50 ± 0.28	69.45 ± 0.35
Foam capacity (%)	40.30 ± 0.28	33.45 ± 0.07
Foam stability (%)	21.70 ± 0.14	18.80 ± 0.15

3.4.3. Emulsifying Properties

Charged or non-charged polar amino acids in proteins can affect the proteins' emulsifying and surfactant properties. Results of emulsifying properties of UE and SE are shown in Table 4. At pH 7.0, the emulsifying properties of UE and SE were significantly different from one another, with values of 45.55 ± 0.64% and 40.25 ± 0.07%, respectively. However, the emulsifying property of UE was significantly different and was higher (85.50 ± 0.28%) than that of SE (69.45 ± 0.35%). The UE had pronounced effects on the emulsifying properties, since it might be exposing more hydrophobic groups to the water and oil interface, giving rise to increased emulsifying capacity and more stable emulsion.

3.4.4. Foaming Properties

Foam formation is similar to emulsion formation. In foam formation, water molecules surround air droplets, which is a nonpolar phase. At pH 7, the foam capacity (FC) of UE was considerably higher than that of SE, with values of 40.30 ± 0.28% and 33.45 ± 0.07%, respectively. The foam stability (FS) of UE was significantly higher than that of SE, with values of 21.70 ± 0.14% and 18.80 ± 0.15%, respectively (Table 4). The results obtained show that the UE, compared to SE, has more flexible protein structure in aqueous solution, and forms more stable foams for stronger interaction on the air–water interface. The variations in foaming properties of UE and SE might be explained with the difference in their protein concentrations. A high protein concentration will improve foam capacity and stability, increasing the viscosity and promoting the formation of the interfacial multilayer membrane [34].

3.5. Scanning Electron Microscopy Analysis

Protein samples obtained using various extraction methods were investigated by the SEM method. Figure 6 shows the surface states of protein samples extracted from whole-body, de-oiled protein (DSF) powder, stirred extraction (SE), and ultrasonic extraction (UE). These studies focused on the study of the surface layer of protein molecules. Accordingly, in Figure 6A, adhesive surface layers of protein molecules in dried DSF can be seen. This state has a collapsed form, which shows the density of protein molecules in relation to one other; therefore, these molecules have a higher crystalline degree than other states, and this will result in lower water solubility. In Figure 6B, on the surface of dried SE protein layers, a plurality of elements similar to "wood shavings" can be seen, in contrast to Figure 6A. These elements formed as a result of separation from the collapsed surface. This means that during drying after SE, from the crystalline state shown in Figure 6A, the protein molecules transition from a crystalline to amorphous state, as shown in Figure 6B; the outcome of this results in an increase in the solubility of the protein molecules in water. Also, in Figure 6C, you can see the results of the drying of protein molecules obtained after UE. This picture clearly shows the increase the number of "wood shavings" on the surface, compared to Figure 6B, and how the transformation collapse forms into thin film pieces. Therefore, UE leads to the formation of scorpion protein molecules from a crystalline state to an amorphous one, which results in an increase of their water solubility.

Figure 6. Scanning electron microscope images of (**A**) DSF, (**B**) SE, and (**C**) UE. (1.00 KX magnifications, bar 10 μm).

4. Conclusions

Studies have shown that the sum of proteins extracted from scorpion bodies had good emulsifying and foaming properties. At the same time, when they were subjected to extraction under different conditions, UE was found to be the best way to obtain the sum of proteins with improved water solubility. According to SEM analysis, the improvement of water solubility of the sum of protein components is accompanied by the transition of these macromolecules from the collapsed and crystalline form to a more amorphous and friable phase. For the extraction of scorpion body parts, some simple ways, i.e., cold water, hot water, and ethanol extraction methods used previously [35,36] were used, and SPs obtained by these methods were physiologically effective for the treatment of various diseases. However, we think that using hot water or ethanol solvents may generally result in the denaturation of native compounds and reduce the real activities of the extracted SPs. Therefore, using UE in an alternative way, at the optimal conditions of UE (i.e., X_1, X_2, and X_3 are 50 min, 400 w, and 18 mL/g, respectively) resulted in increased water solubility and improvement of its emulsifying and foaming properties. In our opinion, the SPs obtained by the UE way should keep a number of their above-mentioned activities, and will be planned for use in creating medicines on that basis in TCM. On the other hand, UE could be the basis for the creation of water-soluble supramolecular complexes for those drugs that are poorly soluble in water. Obtaining such drugs in this way will contribute in the future to (1) an increase the drugs' water solubility, (2) a reduction their therapeutic doses, (c) increase pharmacological action, and (d) reduce toxicity and side effects. Since the sum of proteins from the scorpion body are not individual components, it might be promising, if created in their base gel drug forms, that they could be used through the skin.

Author Contributions: Project administration, A.Y., H.A.A.; Investigation, A.W.(Ahmidin Wali), A.A. and A.O.; Methodology, A.W.(Atikan Wubulikasimu) and Y.G.; Formal analysis, S.M.

Funding: This research was funded by the Program for National Science and Technology Major Project of China, grant number 2017ZX09301045 and "the light of the western" talent training plan: Planting of Hotan Rose and Its Polyphenols Application, Standardized Planting Techniques of Repellent Vernonia and Its Collection, Preservation and Popularization; It is also supported by the PIFI Fund of the Chinese Academy of Sciences for invited scientists, grant number 2019VBA0013 and Central Asian Drug Research and Development Center of the Chinese Academy of Sciences.

Conflicts of Interest: The authors declare no conflict of interest.

References

1. Cologna, C.T.; Silvana, M.; Giglio, J.R.; Soares, A.M.; Arantes, E.C. Tityus serrulatus scorpion venom and toxins: An overview. *Protein Pept. Lett.* **2009**, *16*, 920–932. [CrossRef]

2. Asmari, A.K.A.; Zahrani, A.G.A.; Said, A.J.; Mohammed, A. Clinical aspects and frequency of scorpion stings in the Riyadh Region of Saudi Arabia. *Saudi Med. J.* **2012**, *33*, 852–858. [PubMed]

3. Shi, L.; Zhang, T.X.; Du, C.Y.; Yuan, R.; Wang, C.C.; Li, F. Research Progress on Chemical Constituents, Pharmacological Effects and Clinical Applications of Scorpio. *J. Liaoning Univ. Tradit. Chin. Med.* **2015**, *4*, 31.

4. Yu, M.; Liu, S.L.; Sun, P.B.; Pan, H.; Tian, C.L.; Zhang, L.H. Peptide toxins and small-molecule blockers of BK channels. *Acta Pharmacol. Sin.* **2016**, *37*, 56–66. [CrossRef] [PubMed]

5. Machado, R.J.A.; Estrela, A.B.; Nascimento, A.K.L.; Melo, M.M.A.; Torres-Rêgo, M.; Lima, E.O.; Rocha, H.A.O.; Carvalho, E.; Silva-Junior, A.A.; Fernandes-Pedrosa, M.F. Characterization of TistH, a multifunctional peptide from the scorpion Tityus stigmurus: Structure, cytotoxicity and antimicrobial activity. *Toxicon Off. J. Int. Soc. Toxinol.* **2016**, *119*, 362–370. [CrossRef]

6. Ortiz, E.; Gurrola, G.B.; Schwartz, E.F.; Possani, L.D. Scorpion venom components as potential candidates for drug development. *Toxicon* **2015**, *93*, 125–135. [CrossRef]

7. Melo, E.T.D.; Estrela, A.B.; Santos, E.C.G.; Machado, P.R.L.; Farias, K.J.S.; Torres, T.M.; Carvalho, E.; Lima, J.P.M.S.; Silva-Júnior, A.A.; Barbosa, E.G. Structural characterization of a novel peptide with antimicrobial activity from the venom gland of the scorpion Tityus stigmurus: Stigmurin. *Peptides* **2015**, *68*, 3–10. [CrossRef]

8. Amorim, F.G.; Cordeiro, F.A.; Pinheiro-Júnior, E.L.; Boldrini-França, J.; Arantes, E.C. Microbial production of toxins from the scorpion venom: Properties and applications. *Appl. Microbiol. Biotechnol.* **2018**, *102*, 6319–6331. [CrossRef]

9. Pimenta, A.M.; StöCklin, R.; Favreau, P.; Bougis, P.E.; Martin-Eauclaire, M.F. Moving pieces in a proteomic puzzle: Mass fingerprinting of toxic fractions from the venom of Tityus serrulatus (Scorpiones, Buthidae). *Rapid Commun. Mass Spectrom. Rcm* **2010**, *15*, 1562–1572. [CrossRef]

10. Shao, J.; Rong, Z.; Xin, G.; Yang, B.; Zhang, J. Analgesic Peptides in Buthus martensii Karsch: A Traditional Chinese Animal Medicine The Analgesic Peptides in Scorpion Venoms. *Asian J. Tradit. Med.* **2007**, *2*, 45–50.

11. Pei, W.; Jie, T.; Yi, Z. Recombinant Expression and Functional Characterization of Martentoxin: A Selective Inhibitor for BK Channel ($\alpha + \beta 4$). *Toxins* **2014**, *6*, 1419–1433.

12. Goudet, C.; Chi, C.W.; Tytgat, J. An overview of toxins and genes from the venom of the Asian scorpion Buthus martensi Karsch. *Toxicon* **2002**, *40*, 1239–1258. [CrossRef]

13. Gupta, S.D.; Debnath, A.; Saha, A.; Giri, B.; Tripathi, G.; Vedasiromoni, J.R.; Gomes, A.; Gomes, A. Indian black scorpion (Heterometrus bengalensis Koch) venom induced antiproliferative and apoptogenic activity against human leukemic cell lines U937 and K562. *Leuk. Res.* **2007**, *31*, 817–825. [CrossRef] [PubMed]

14. Díaz-García, A.; Morier-Díaz, L.; Frión-Herrera, Y.; Rodríguez-Sánchez, H.; Caballero-Lorenzo, Y.; Mendoza-Llanes, D.; Riquenes-Garlobo, Y.; Fraga-Castro, J.A. In vitro anticancer effect of venom from Cuban scorpion Rhopalurus junceus against a panel of human cancer cell lines. *J. Venom Res.* **2013**, *4*, 5–12.

15. Guo, X.; Ma, C.; Du, Q.; Wei, R.; Wang, L.; Zhou, M.; Chen, T.; Shaw, C. Two peptides, TsAP-1 and TsAP-2, from the venom of the Brazilian yellow scorpion, Tityus serrulatus: Evaluation of their antimicrobial and anticancer activities. *Biochimie* **2013**, *95*, 1784–1794. [CrossRef]

16. He, Y.; Zhao, R.; Di, Z.; Li, Z.; Xu, X.; Hong, W.; Wu, Y.; Zhao, H.; Li, W.; Cao, Z. Molecular diversity of Chaerilidae venom peptides reveals the dynamic evolution of scorpion venom components from Buthidae to non-Buthidae. *J. Proteom.* **2013**, *89*, 1–14. [CrossRef]

17. Ali, S.A.; Alam, M.; Abbasi, A.; Kalbacher, H.; Schaechinger, T.J.; Hu, Y.; Zhijian, C.; Li, W.; Voelter, W. Structure–Activity Relationship of a Highly Selective Peptidyl Inhibitor of Kv1.3 Voltage-Gated K^+-Channel from Scorpion (B. sindicus) Venom. *Int. J. Pept. Res.* **2014**, *20*, 19–32. [CrossRef]

18. Hoang, N.A.; Maksimov, G.V.; Berezin, B.B.; Piskarev, V.E.; Yamskov, I.A. Isolation of toxins from buthus occitanus sp. scorpion and their action on excitability of myelinated nerve. *Bull. Exp. Biol. Med.* **2001**, *132*, 953–955. [CrossRef]

19. Hoang, N.A.; Berezin, B.B.; Lakhtin, V.M.; Yamskov, I.A. Isolation and partial characterization of lectin from the venom of vietnamese scorpion buthus occitanussp. *Appl. Biochem. Microbiol.* **2001**, *37*, 534–537. [CrossRef]

20. Ren, Y.; Wu, H.; Lai, F.; Yang, M.; Li, X.; Tang, Y. Isolation and identification of a novel anticoagulant peptide from enzymatic hydrolysates of scorpion (buthus martensii karsch) protein. *Food Res. Int.* **2014**, *64*, 931–938. [CrossRef]

21. Ma, T.; Zhu, H.; Jing, W.; Qiang, W.; Yu, L.; Sun, B. Influence of extraction and solubilizing treatments on the molecular structure and functional properties of peanut protein. *Lwt Food Sci. Technol.* **2017**, *79*, 197–204. [CrossRef]

22. Ahmedna, M.; Prinyawiwatkul, W.; Rao, R.M. Solubilized wheat protein isolate: Functional properties and potential food applications. *J. Agric. Food Chem.* **1999**, *47*, 1340–1345. [CrossRef] [PubMed]

23. Zhang, Y.; Zhang, J.; Sheng, W.; Wang, S.; Fu, T.J. Effects of heat and high-pressure treatments on the solubility and immunoreactivity of almond proteins. *Food Chem.* **2015**, *199*, 856–861. [CrossRef] [PubMed]

24. Thaiphanit, S.; Anprung, P. Physicochemical and emulsion properties of edible protein concentrate from coconut (Cocos nucifera L.) processing by-products and the influence of heat treatment. *Food Hydrocoll.* **2016**, *52*, 756–765. [CrossRef]

25. Feng, X.L.; Liu, H.Z.; Shi, A.M.; Liu, L.; Wang, Q.; Adhikari, B. Effects of transglutaminase catalyzed crosslinking on physicochemical characteristics of arachin and conarachin-rich peanut protein fractions. *Food Res. Int.* **2014**, *62*, 84–90. [CrossRef]

26. Jeong, H.J.; Park, J.H.; Lam, Y.; de Lumen, B.O. Characterization of lunasin isolated from soybean. *J. Agric. Food Chem.* **2003**, *51*, 7901–7906. [CrossRef]

27. Lin, X.S.; Wen, Q.; Huang, Z.L.; Cai, Y.Z.; Halling, P.J.; Yang, Z. Impacts of ionic liquids on enzymatic synthesis of glucose laurate and optimization with superior productivity by response surface methodology. *Process. Biochem.* **2015**, *50*, 1852–1858. [CrossRef]

28. Mao, X.; Hua, Y. Composition, Structure and Functional Properties of Protein Concentrates and Isolates Produced from Walnut (Juglans regia L.). *Int. J. Mol. Sci.* **2012**, *13*, 1561–1581. [CrossRef]

29. Arcan, I.; Yemenicioğlu, A. Effects of controlled pepsin hydrolysis on antioxidant potential and fractional changes of chickpea proteins. *Food Res. Int.* **2010**, *43*, 140–147. [CrossRef]

30. Chityala, P.K. Effect of xanthan/enzyme-modified guar gum mixtures on the stability of whey protein isolate stabilized fish oil-in-water emulsions. *Food Chem.* **2016**, *212*, 332–340. [CrossRef]

31. Gong, K.J.; Shi, A.M.; Liu, H.Z.; Liu, L.; Hu, H.; Adhikari, B.; Wang, Q. Emulsifying properties and structure changes of spray and freeze-dried peanut protein isolate. *J. Food Eng.* **2016**, *170*, 33–40. [CrossRef]

32. Wu, H.; Wang, Q.; Ma, T.; Ren, J. Comparative studies on the functional properties of various protein concentrate preparations of peanut protein. *Food Res. Int.* **2009**, *42*, 343–348. [CrossRef]

33. Jianmei, Y.; Mohamed, A.; Ipek, G. Peanut protein concentrate: Production and functional properties as affected by processing. *Food Chem.* **2007**, *103*, 121–129.

34. Kadam, S.U.; Tiwari, B.K.; Álvarez, C.; O'Donnell, C.P. Ultrasound applications for the extraction, identification and delivery of food proteins and bioactive peptides. *Trends Food Sci. Technol.* **2015**, *46*, 60–67. [CrossRef]

35. Liu, T.H.; Zhang, C.X.; Zhuang, F.J.; Xu, X.P.; Chi, Y.S. A Review on the Chemical Components and the Edible and Medicinal Value of Scorpion. *Hans. J. Food Nutr. Sci.* **2017**, *6*, 165–174. [CrossRef]

36. Wu, F.L.; Dong, Q.H.; Wang, H.; Tan, J.; Lin, H.Q.; Li, P.Y.; Li, Y.J. Research Progress of Chinese Scorpion. *J. Liaoning Univ. TCM* **2018**, *20*, 108–111.

Sample Availability: Samples of the compounds are available from the authors.

 *molecules*

Article

Influence of Potato Crisps Processing Parameters on Acrylamide Formation and Bioaccesibility

Emmanuel Martinez [1], Jose A. Rodriguez [1], Alicia C. Mondragon [2], Jose Manuel Lorenzo [3] and Eva M. Santos [1,*]

[1] Area Academica de Quimica, Universidad Autonoma del Estado de Hidalgo, Carr. Pachuca-Tulancingo Km. 4.5, Mineral de la Reforma, 42184 Hidalgo, Mexico; qaemmanuel@gmail.com (E.M.); josear@uaeh.edu.mx (J.A.R.)

[2] Laboratorio de Higiene, Inspeccion y Control de Alimentos, Departamento de Quimica Analitica, Nutricion y Bromatologia, Facultad de Veterinaria, Universidad de Santiago de Compostela, 27002 Lugo, Spain; alicia.mondragon@deinal.es

[3] Meat Technology Centre of Galicia, Rúa Galicia N° 4, Parque Tecnológico de Galicia, San Cibrao das Viñas, 32900 Ourense, Spain; jmlorenzo@ceteca.net

* Correspondence: emsantos@uaeh.edu.mx

Academic Editor: Jose M. Miranda
Received: 24 September 2019; Accepted: 21 October 2019; Published: 23 October 2019

Abstract: A fractional factorial design was used to evaluate the effects of temperature, frying time, blanching treatment and the thickness of potato slices on acrylamide content in crisps. The design was used on freshly harvested and four-month stored potatoes. The critical factors found were temperature and frying time, and the interaction between blanching treatment and slice thickness. Once frying conditions were selected, an acrylamide content of 725 and 1030 mg kg^{-1} was found for non-stored and 4-month stored tubers, with adequate textural parameters in both cases. The difference in concentration is related to storage conditions, which must be controlled in order to control acrylamide levels. Bioaccesibility studies demonstrated that acrylamide concentration remained at 70%, and reductions took place mainly at the intestinal phase, as a result of reaction with nucleophilic compounds.

Keywords: acrylamide; crisps; temperature; frying time; blanching; thickness

1. Introduction

Acrylamide, a thermal processing contaminant with a low molecular weight, which is soluble in water, is formed when carbohydrate-rich foods are subjected to temperatures above 120 °C in low-moisture conditions, such as frying, roasting or baking. Several studies consider Maillard reactions to be the main pathway for acrylamide, 5 hydroxymethylfurfural, methylglyoxal–lysine dimers, Nε–carboxymethyl–lysine and pyrraline formation in processed foods, particularly reactions between the carbonyl group of reducing sugars and amino acids [1,2]. Initial reports showed relatively high concentrations of acrylamide in high-carbohydrate foodstuffs, such as crispy bread, breakfast cereals, pastries, coffee, French fries and crisps. In general, potato products present higher acrylamide contents (250–4000 µg kg^{-1}) compared to other food products, because of the higher concentration of asparagine in potato tubers [3,4].

Acrylamide has been classified by the Agency for Research on Cancer as "probably carcinogenic to humans" (Group 2A) and, in the latest report from the European Food Safety Authority Expert Panel on Contaminants in the Food Chain (CONTAM, 2015), the margins of exposure for acrylamide lead to concerns regarding their neoplastic effects, based on animal evidence [5]. Therefore, the European Commission recently established "indicative" levels for the presence of acrylamide in food, suggesting a limit of 750 µg kg^{-1} for crisps [6]. Previously, the European Union and FDA had encouraged food industries to reduce the presence of this contaminant.

In response to this, the food industry has been forced to apply different strategies to reduce acrylamide formation, applying processing modifications according to recommendations published by the European Food and Drink Federation in the document "Acrylamide Toolbox", which includes the following variables: temperature and time during frying, blanching treatment, and thickness of potato slices [7]. However, the main challenge for the food industries is in producing potato chips with low acrylamide levels without affecting their sensory properties [8].

Although it is important to evaluate the effect of pre-treatments on the frying process on acrylamide mitigation, it is also necessary to determine their influence on its bioavailability. Potato chips are an important part of the snack food market in many countries, and few studies have been published about the bioavailability of acrylamide after crisps have been through the digestion process [8,9]. Acrylamide intake in humans occurs mainly via food ingestion, where the total amount of this compound does not necessarily reflect the available amount to the body. There are chemical changes in the food when it enters the digestive tract, because of pH variations, and the action of several enzymes in every stage of the digestion process [10–12]. In this sense, bioaccessibility is used to evaluate the amount of a chemical compound available for absorption after it is released from the food matrix into the gastrointestinal tract. Several methodologies can be used to assess the bioaccessibility of contaminants or nutrients: 1) in vitro, 2) ex vivo, 3) in situ and 4) in vivo models [13,14]. Among the techniques mentioned before, in vitro models are more commonly used than in vivo models, because of their simplicity, lower cost, lack of ethical issues and good reproducibility under controlled conditions.

The present work evaluated some acrylamide mitigation strategies, employing a fractional factorial design experiment, and their influence on textural properties and bioaccesibility, by an in vitro simulated digestion assay.

2. Results and Discussion

2.1. Acrylamide Content

Potatoes of the Atlantic variety were selected for crisp production, because of their low levels of glucose and fructose compared to other varieties, such as the Ranger Russet [15]. According to the storage period of potato tubers, two sampling points were considered: freshly harvested potatoes (summer of 2018) and four-month stored potatoes. The main variables described for acrylamide reduction in potato crisps are [7]: temperature control and time of frying process, blanching, and the thickness of slices. These variables were selected because they are easily applicable in the industry and do not represent extra costs in production. A fractional factorial design (2^{4-1}) was proposed, to evaluate the influence of frying process parameters in the total amount of acrylamide formed and its bioaccessibility. The levels selected for the variables were: temperature (T) of 150 and 180 °C; frying time (t) of 5 and 10 min; blanching treatment (B)—no blanching treatment and 70 °C for 5 min—and thickness of slice (w), at 1.5 and 2.0 mm. In this sense, eight experiments (in duplicate, a total of 16 experiments) were performed.

Table 1 shows the matrix used and the acrylamide content for the experimental design. According to the obtained results, the concentration of acrylamide was higher in stored potatoes. This can be explained by the difference between the content of reducing sugars between samples, since a concentration of 23 mg kg^{-1} of reducing sugars was found in freshly harvested potatoes, while, in samples stored for 4 months, a concentration of 34 mg kg^{-1} was found. During storage, a phenomenon called cold sweetening occurs, consisting of a degradation of the potato starch caused by the low temperatures [16]. Even if appropriate storage conditions of potato tubers are followed, a slight increase of reducing sugars is expected, and will affect the acrylamide concentration [17].

Table 1. Matrix of the experimental fractional factorial design and the average of obtained responses.

Exp	T (°C)	t (min)	B (70 °C, min)	w (mm)	Acrylamide Content (µg kg^{-1}) ± sd	
					Non-stored	*4-month stored*
1	150(−1)	5(−1)	0(−1)	1.5(−1)	973 ± 71	1257 ± 171
2	150(−1)	10(+1)	0(−1)	2.0(+1)	752 ± 63	890 ± 52
3	150(−1)	10(+1)	5(+1)	1.5(−1)	825 ± 67	1483 ± 58
4	150(−1)	5(−1)	5(+1)	2.0(+1)	885 ± 62	1097 ± 205
5	180(+1)	10(+1)	0(−1)	1.5(−1)	2590 ± 243	4386 ± 270
6	180(+1)	5(−1)	0(−1)	2.0(+1)	795 ± 92	1149 ± 82
7	180(+1)	5(−1)	5(+1)	1.5(−1)	1138 ± 220	1812 ± 70
8	180(+1)	10(+1)	5(+1)	2.0(+1)	2250 ± 204	3540 ± 254

A comparison of each variance with respect to the variance of the residual allows the determination of the the the experimental F value (F_{exp}), which is then compared with the $F_{critical}$ value (5.32, $p = 0.05$, 1,8). The F-test indicated that, in both cases (non-stored and stored tubers), the critical factors for acrylamide formation were ($p = 0.05$): temperature (T) and frying time (t). The contribution of potato thickness (w) was also observed for stored samples (Table 2).

Table 2. Analysis of variance of the results obtained from the fractional factorial design, showing the factors significantly affecting acrylamide concentration.

Factor	Non-Stored			4-month Stored		
	Effect	Variance	F_{exp}	Effect	Variance	F_{exp}
T	834.5	2,785,561	**63.72**	1540	9,486,400	**216.99**
t	656.5	1,723,969	**39.43**	1246	6,210,064	**142.05**
B	−3	36	8.2×10^{-4}	62.5	15,625	0.36
w	−211	178,084	4.07	−565.5	1,279,161	**29.26**
Residual		43,718			56,593.5	

Figure 1 shows the mean effect graph for the effect of evaluated control factors on acrylamide concentration. Independently of storage time, the mean effect profile is similar in both cases, demonstrating that the increase in acrylamide concentration is mainly a consequence of the increment of reducing sugars concentration, rather than the processing variables. Bertuzzi et al. also reported a strong influence of reducing sugar concentration on acrylamide formation in processed potato products [18]. A percentage of reducing sugars between 0.15%–0.20% has been suggested as an indicator of the suitability of potatoes for processing [19]. Despite the lower reducing sugar content found in potatoes in this study, acrylamide formation reached values over the 750 µg kg^{-1} limit established by the European Commission [6].

The results obtained are congruent with others previously described, in which an increment in acrylamide content in potatoes, in relation to temperature and frying time, was observed [18,20]. In general, the increase in acrylamide content follows a linear function over time, while the relationship with frying temperature is not linear, although temperatures over 175 °C significantly increased acrylamide levels [1,18,20–22]. Any temperature–time combination proposed to mitigate acrylamide content should not affect the overall quality of the fried product, especially its texture.

Although the other processing parameters (blanching and thickness) had a smaller contribution, their interaction was also significant (Figure 2). As acrylamide is formed on the potato surface, the size:volume ratio of the potato slice influences the acrylamide content. In general, thinner and smaller cut sizes result in increased acrylamide formation [22]. Also, the blanching of potato slides prior to the frying process has been recommended to reduce the acrylamide content, since this helps remove reducing sugars on the surface [8,23]. In this study, a blanching process was required for thinner potatoes, to decrease acrylamide formation. However, a lower concentration of acrylamide was observed in thicker potatoes without the blanching treatment. It seems that, in thicker slices, heat diffusion is slower when the slice has not been blanched and, consequently, a decrement of acrylamide content can be observed. So, the blanching treatment should not be suitable for potato slices with a thickness over 2.00 mm. This result is interesting for the industry because the application of blanching treatments at high temperatures and longer times represents an additional difficulty, since most industrial frying processes are continuous, and blanching could be unnecessary for certain types of crisps.

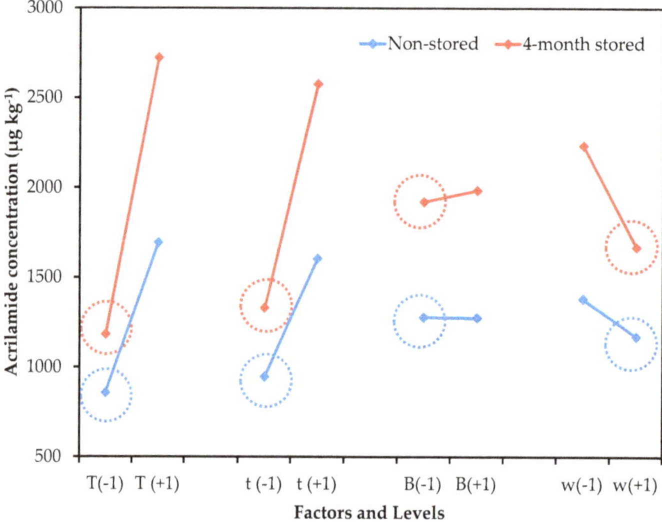

Figure 1. Mean effect graph vs. acrylamide concentration ($\mu g\ kg^{-1}$) of the experimental design for non-stored and 4-month stored potato tubers storage.

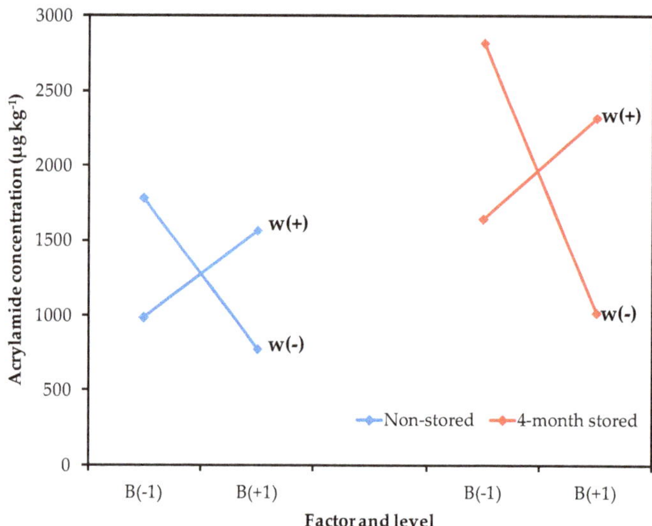

Figure 2. Mean effect graph vs. acrylamide concentration ($\mu g\ kg^{-1}$) for the interaction between blanching (B) and thickness (w) in non-stored and 4-month stored tubers.

The combination of settings that generated the minimum acrylamide concentration was: temperature, 150 °C (−1); time, 5 min (−1); no blanching treatment (−1); and a slice thickness of 2.0 mm (+1). Acrylamide content of 725 and 1030 mg kg^{-1} was determined at these conditions for non-stored and 4-month stored tubers, respectively.

2.2. Texture Analysis and Bioaccesibility

Potato chips are consumed as indulgent foods, with a characteristic flavor and crispiness, the latter evaluated through textural parameters. In order to evaluate it, crisps were prepared according to the processing parameters detailed above. Figure 3 shows a representative profile of the force (N) vs. probe displacement (mm). The force displacement curves have a jagged appearance, with several fracture events, which are typical of crispy food [24]. These graphics have two well-differentiated regions, the first one starting from the first contact between the potato crisp and the probe, until the major force is achieved (associated with major structural breakdown and hardness). The second region starts from the major structural breakdown and continues until the end, where smaller force events take place. The parameters evaluated were the maximum force applied (N), related to hardness or firmness, and the gradient (N s^{-1}), which is related to stiffness.

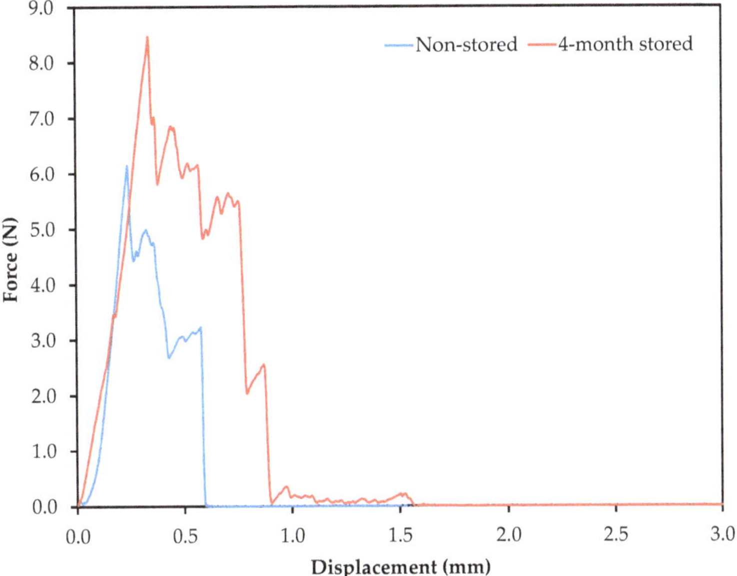

Figure 3. Force displacement curves generated from the analysis of potato crisps obtained from frying non-stored and 4-month stored tubers. Frying conditions: temperature, 150 °C; time, 5 min; no blanching treatment; and a slice thickness of 2.0 mm.

The maximum force applied was 6.1 and 8.5 N, for non-stored and 4-month stored samples, respectively. Genovese et al. have reported values below 6 N for firmness [25]. The storage of the tubers promoted an increase in firmness. On the other hand, samples obtained from frying the non-stored potato slices presented the highest gradient (8.9 N s^{-1}) compared to 4-month stored samples (7.2 N s^{-1}), making them, therefore, the stiffest. Additionally, the number of total force peaks in each case can be evaluated: seven for non-stored tubers and 13 for 4-month stored tubers. Taking into account that values higher than six are considered high-sensory crispiness, both samples can be accepted as adequate snacks [26].

Once the effect of the control factors on acrylamide formation had been evaluated, it was important to know their effect on bioaccesibility. Values of 79.1% ± 1.5% and 76.9% ± 1.5% of bioaccesibility were obtained for the snacks using potatoes that were freshly harvested and stored for 4 months before frying, respectively. The results obtained in this study are in agreement with the predicted behavior of potato [10] and bakery products [12]. The main effect was observed at the end of the intestinal phase of the digestion process, and could be explained by the formation of Michael adducts, in which the acrylamide would be involved.

During the gastric phase of the digestion process, pepsin hydrolyzes the proteins present in the potato, generating small peptide chains and some amino acid residues, such as cysteine and lysine, that have a nucleophilic character (-SH and -NH$_2$). These peptides are able to interact with acrylamide and form adducts, which consequently causes an apparent reduction in acrylamide during the intestinal phase [12]. Taking into account the mentioned results, this can be proposed as an alternative method of reducing acrylamide concentration in the digestion process, through the formation of adducts with thiol compounds (i.e., glutathione) contained in other foods, such as spinach, avocados or asparagus [27,28]

3. Materials and Methods

3.1. Reagents

Acrylamide, 2-naphthalenethiol, and the following enzymes: pepsin ($\geq$250 U mg^{-1} solid), from porcine gastric mucosa, pancreatin (4 × USP), from porcine pancreas, α-amylase, from human saliva (500 U) and porcine bile salts, were purchased from Sigma–Aldrich (St. Louis, MO, USA). Potassium chloride, sodium bicarbonate and sodium chloride were purchased from MEYER (Mexico City, Mexico). Acetonitrile (HPLC grade) and acetic acid were obtained from J.T. Baker (Philisburg, NJ, USA).

3.2. Samples and Frying Conditions

Potato (Atlantic variety) and vegetable oil (palm olein) were the raw materials used, provided by Fritos Totis (Tizayuca, Hidalgo, Mexico). Samples were collected in two periods, the first with potatos freshly harvested and the second after four months of storage, at 8 °C and 95% relative humidity. Slices (with a thickness of 1.5 and 2.0 mm, and a diameter of around 35 mm) were cut using a slicing machine. Slices were rinsed immediately after cutting for 1 min in deionized water (resistivity of 18.2 MΩ cm), to eliminate some starch material adhered onto the surface, before frying. Forty grams of slides were deep-fried in an electrical fryer Blazer FE-3 (Mexico city, Mexico) 1600 W, 127 V and equipped with a bowl with 3 L of capacity, a static basket and a regulating thermometer.

The main variables evaluated for acrylamide reduction in potato crisps were [6]: control of temperature and time of frying process, blanching, and slice thickness. These strategies were selected because they are easily applicable to the industry and do not represent extra costs in production. A fractional factorial design (2^{4-1}) was proposed, to evaluate the influence of frying process parameters in the total amount of acrylamide formed and its bioaccessibility. The factors selected were temperature (T), frying time (t), blanching treatment (B), and thickness of potato slices (w). Results obtained from the fractional factorial design were subjected to an analysis of variance study. The mean effect of each factor on acrylamide content and the variance was estimated through the Yates algorithm [29]. Experimental results were analyzed using MINITAB® version 17 software (Minitab Inc., State College, PA, USA).

3.3. Acrylamide Determination

Acrylamide determination was carried out by chemical derivatization, using 2-naphthalenethiol as a derivatization reagent, followed by HPLC separation and quantification with fluorescence detection [30]. In brief, 3 g of potato crisps were homogenized with 30 mL of deionized water and then defatted with hexane. Then, the aqueous phase was centrifuged and filtered using a cellulose membrane (pore size, 0.45 μm). Afterwards, an aliquot of 7.5 mL was mixed with 1 mL of the solution

2-naphthalenethiol (62 mM), in a solution of sodium hydroxide (0.1 M). The mixture was heated at 90 °C for 45 min. The reaction was stopped with 1.5 mL of acetic acid (1%) and the solution was centrifuged at 700 g for 15 min. Acrylamide concentration of samples was calculated by standard additions, by spiking each sample with known concentrations of acrylamide.

Chromatography separation was performed using an Agilent Technologies 1260 Infinity chromatography system (Agilent Technologies, Waldbronn, Germany). Samples were manually injected using a 20-μL loop, and a C-8 ZORBAX eclipse XDB column (5 μm; 150 × 4.6 mm internal diameter) from Agilent Technologies was used as the stationary phase. The mobile phase was acetic acid (1.0% *v/v*) and acetonitrile in a 50:50 ratio (*v/v*). The flow rate was 0.8 mL min^{-1}.

3.4. Texture Evaluation

A TA.XTPlus Texture Analyser (Stable Micro Systems Ltd., Surrey, UK), equipped with a 5 kg load cell, was used for force/displacement measurements, with a stainless spherical probe (P/0.25S) of a 0.25 inch diameter. The samples were placed on a crisp fracture rig (HDP/CFS). The test settings were: test speed of 1 mm s^{-1}, trigger force of 0.406 N, travel distance of 3 mm.

3.5. In Vitro Digestion

Bioaccessibility was estimated using an in vitro digestion model (oral, gastric and intestinal stages) and the simulated fluids—salivary (SFS), gastric (SFG) and intestinal (SFI)—were prepared according to an internationally agreed protocol (Table 3) [31]. Samples of potato crisps (5 g) were transferred to a conical tube and mixed with 5 mL SFS (containing 75 U α-amylase mL^{-1}) for 3 min. After that, 5 mL of a pepsin solution (12.5 mg mL^{-1} in 0.1 M HCl) and 10 mL of SFG were added. The mixture was adjusted to pH 2.0 and incubated in a shaking water bath at 37 °C for 2 h, with an agitation speed of 60 strokes per min. After incubation, the pH was adjusted to 7.5, and then bile salts (10 mg mL^{-1}), dissolved in 20 mL of SFI together with a pancreatin solution (5 mL, 10 mg mL^{-1} in water), were added to the tube. The mixture was incubated at 37 °C for 2 h, with shaking at an agitation speed of 60 strokes per min, to simulate the duodenal phase. Digestion products were placed in a dialysis bag (6–8000 molecular weight cut off; Sigma Aldrich) and dialyzed in 250 mL of sodium bicarbonate solution (pH = 7.5) for 12 h. Dialysis aliquots were removed and acrylamide dialyzed was determined by the method described above. Bioaccesibility was reported as the percentage of acrylamide in the dialyzed sample compared to the value in the original sample.

Table 3. Composition of simulated salivary, gastric and intestinal electrolytes fluids.

Compound	Simulated Salivary Fluid (SSF) mmol L^{-1}	Simulated Gastric Fluid (SGF) mmol L^{-1}	Simulated Intestinal Fluid (SIF) mmol L^{-1}
KCl	15.10	6.90	6.80
KH$_2$PO$_4$	3.70	0.90	0.80
NaHCO$_3$	13.60	25.00	85.00
NaCl	-	47.20	38.40
MgCl$_2$·6H$_2$O	0.15	0.10	0.33
(NH$_4$)$_2$CO$_3$	0.06	0.50	-
CaCl$_2$·2H$_2$O	1.50	0.15	0.60

4. Conclusions

The evaluation of potato crisps' processing parameters was performed by a chemometric approach, through fractional factorial design. The factors with a higher contribution were temperature and frying time, and an increment on the acrylamide, with respect to the store time of raw material, was also observed. Crisps obtained in the most suitable conditions have adequate texture parameters, while bioaccesibility takes place mainly at the intestinal phase, through reaction with –NH$_2$ and

–SH containing compounds. In order to reduce acrylamide concentration in the digestion process, the consumption of foods which contain compounds with these functional groups is proposed.

Author Contributions: E.M., J.A.R. and E.M.S. performed the experiments; E.M. and A.C.M. performed bioaccesibility experiments; J.M.L. analyzed the texture data. J.A.R. and E.M.S. are responsible for the writing of the work.

Funding: This research was funded by Fritos Totis SA de CV and CONACyT-Incentives for Research, Technological Development and Innovation Program (conv 2016-232271).

Acknowledgments: Programa para el Desarrollo Profesional Docente (PRODEP), Consejo Nacional de Ciencia y Tecnología (SNI membership), and Programa de Fortalecimiento de la Calidad Educativa (PFCE-2019).

Conflicts of Interest: The authors declare no conflict of interest.

References

1. Gökmen, V. *Acrylamide in Food: Analysis, Content and Potential Health Effects*; Academic Press: San Diego, CA, USA, 2016; pp. 1–9.

2. Oh, M.J.; Kim, Y.; Lee, S.H.; Lee, K.W.; Park, H.Y. Prediction of CML contents in the Maillard reaction products for casein-monosaccharides model. *Food Chem.* **2018**, *267*, 271–276. [CrossRef] [PubMed]

3. Blank, I. Current status of acrylamide research in food: Measurement, safety, assessment, and formation. *Annals New York Academy Sci.* **2015**, *1043*, 30–40. [CrossRef] [PubMed]

4. Bethke, P.C.; Bussan, A. Acrylamide in processed potato products. *Am. J. Potato Res.* **2013**, *90*, 403–424. [CrossRef]

5. Agents Classified by the IARC Monographs, Volumes 1–123. Available online: https://monographs.iarc.fr/agents-classified-by-the-iarc/ (accessed on 20 September 2019).

6. EC (European Commission). Commission Regulation EU of 20 November 2017 Establishing Mitigation Measures and Benchmark Levels for the Reduction of the Presence of Acrylamide in Food. Available online: https://publications.europa.eu/en/publication-detail/-/publication/1f3b45fb-ce6b-11e7-a5d5-01aa75ed71a1/language-en (accessed on 20 September 2019).

7. Food Drink Europe. Acrylamide Toolbox. Available online: https://www.fooddrinkeurope.eu/publication/fooddrinkeurope-updates-industry-wide-acrylamide-toolbox/ (accessed on 20 September 2019).

8. Pedreschi, F.; Mariotti, S.; Granby, K.; Risum, J.R. Acrylamide reduction in potato chips by using commercial asparaginase in combination with conventional blanching. *LWT-Food Sci. Technol.* **2011**, *44*, 1473–1476. [CrossRef]

9. Eriksson, S.; Karlsson, P. Alternative extraction techniques for analysis of acrylamide in food: Influence of pH and digestive enzymes. *LWT-Food Sci. Technol.* **2006**, *39*, 393–399. [CrossRef]

10. Sansano, M.; Heredia, A.; Peinado, I.; Andres, A. Dietary acrylamide: What happens during digestion. *Food Chem.* **2017**, *237*, 58–64. [CrossRef]

11. Moreda-Piñeiro, J.; Moreda-Piñeiro, A.; Romarís-Hortas, V.; Moscoso-Perez, C.; López-Mahía, P.; Muniategui-Lorenzo, S.; Bermejo-Barrera, P.; Prada-Rodríguez, D. In-vivo and in-vitro testing to assess the bioaccessibility and the bioavailability of arsenic, selenium and mercury species in food samples. *Trends Anal. Chem.* **2011**, *30*, 324–345. [CrossRef]

12. Hamzalioğlu, A.; Gökmen, V. Investigation of the reactions of acrylamide during in vitro multistep enzymatic digestion of thermally processed foods. *Food Funct.* **2014**, *6*, 109–114. [CrossRef]

13. Carbonell-Capella, J.M.; Buniowska, M.; Barba, F.J.; Esteve, M.J.; Frigola, A. Analytical methods for determining bioavailability and bioaccessibility of bioactive compounds from fruits and vegetables: A review. *Comprehens. Rev. Food Sci. Food Safety* **2014**, *13*, 155–171. [CrossRef]

14. Barba, J.F.; Mariutti, R.B.L.; Bragagnolo, N.; Mercadante, Z.A.; Barbosa-Canovas, G.V.; Orlien, V. Review: Bioaccessibility of bioactive compounds from fruits and vegetables after thermal and nonthermal processing. *Trend. Food Sci. Technol.* **2017**, *67*, 195–206. [CrossRef]

15. Rommens, C.M.; Yan, H.; Swords, K.; Richael, C.; Ye, J. Low-acrylamide French fries and potato chips. *Plant Biotechnol. J.* **2008**, *6*, 843–853. [CrossRef] [PubMed]

16. Sowokinos, J. The molecular and cellular biology of the potato. In *Stress-induced Alterations in Carbohydrate Metabolism*; Vayda, M.E., Park, W.D., Eds.; CAB international: Wallingford, UK, 1990; pp. 137–157.

17. Medeiros, V.R.; Mestdagh, F.; Van Poucke, C.; Van Peteghem, C.; De Meulenaer, B. A two-year investigation towards and effective quality control of incoming potatoes as an acrylamide mitigation strategy in french fries. *Food Add. Contaminants* **2012**, *29*, 362–370.

18. Bertuzzi, T.; Mulazzi, A.; Rastelli, S.; Sala, L.; Pietri, A. Mitigation measures for acrylamide reduction in dough-based potato snacks during their expansion by frying. *Food Add. Contaminants Part A* **2018**, *35*, 1940–1947. [CrossRef] [PubMed]

19. Pedreschi, F.; Mariotti, M.S.; Granby, K. Current issues in dietary acrylamide: Formation, mitigation and risk assessment. *J. Sci. Food Agricul.* **2014**, *94*, 9–20. [CrossRef]

20. Pedreschi, F.; Moyano, P.; Kaack, K.; Granby, K. Color changes and acrylamide formation in fried potato slices. *Food Res. Int.* **2005**, *38*, 1–9. [CrossRef]

21. Yang, Y.; Achaerandio, I.; Pujola, M. Influence of the frying process and potato cultivar on acrylamide formation in French fries. *Food control* **2016**, *62*, 216–223. [CrossRef]

22. Vinci, R.M.; Mestdagh, F.; De Meulenaer, B. Acrylamide formation in fried potato products–Present and future, a critical review on mitigation strategies. *Food Chem.* **2012**, *133*, 1138–1154. [CrossRef]

23. Abboudi, M.; Al-Bachir, M.; Koudsi, Y.; Jouhara, H. Combined effects of gamma irradiation and blanching process on acrylamide content in fried potato strips. *Int. J. Food Propert.* **2016**, *19*, 1447–1454. [CrossRef]

24. Chen, J.; Karlsson, C.; Povey, M. Acoustic envelope detector for crispness assessment of biscuits. *J. Texture Studies* **2005**, *36*, 139–156. [CrossRef]

25. Genovese, J.; Tappi, S.; Luo, W.; Tylewicz, U.; Marzocchi, S.; Marziali, S.; Romani, S.; Ragni, L.; Rocculi, P. Important factors to consider for acrylamide mitigation in potato crisps using pulsed electric fields. *Innovative Food Sci. Emerging Technol.* **2019**, *55*, 18–26. [CrossRef]

26. Salvador, A.; Varela, P.; Sanz, T.; Fiszman, S.M. Understanding potato chips crispy texture by simultaneous fracture and acoustic measurements, and sensory analysis. *LWT-Food Sci. Technol.* **2009**, *42*, 763–767. [CrossRef]

27. Tong, G.C.; Cornwell, W.K.; Means, G.E. Reactions of acrylamide with glutathione and serum albumin. *Toxicol. Lett.* **2004**, *147*, 127–131. [CrossRef] [PubMed]

28. Ghareeb, D.A.; Khalil, A.A.; Elbassoumy, A.M.; Hussien, H.M.; Abo-Sraiaa, M.M. Ameliorated effects of garlic (Allium sativum) on biomarkers of subchronic acrylamide hepatotoxicity and brain toxicity in rats. *Toxicol. Envir. Chem.* **2010**, *92*, 1357–1372. [CrossRef]

29. Massart, D.L.; Vandeginste, B.G.M.; Buydens, L.M.C.; De Jong, S.; Lewi, P.J.; Smeyers-Verbeke, J. *Chemometrics: A Textbook*, 1st ed.; Elsevier: Amsterdam, The Netherlands, 1988; pp. 659–682.

30. Martinez, E.; Rodriguez, J.A.; Bautista, M.; Rangel-Vargas, E.; Santos, E.M. Use of 2-Naphthalenethiol for derivatization and determination of acrylamide in potato crisps by high-performance liquid chromatographic with fluorescence detection. *Food Analy. Methods* **2018**, *11*, 1636–1644. [CrossRef]

31. Minekus, M.; Alminger, M.; Alvito, P.; Balance, S.; Bohn, T.; Bourlieu, C.; Carriere, F.; Boutrou, R.; Corredig, M.; Dupont, D.; et al. A standardised static in-vitro digestion method suitable for food—An international consensus. *Food Function* **2014**, *5*, 1113–1124. [CrossRef]

Sample Availability: Samples of the compounds are not available from the authors.

Article

Determination of *N*-Carbamylglutamate in Feeds and Animal Products by High Performance Liquid Chromatography Tandem Mass Spectrometry

Yonghang Ma, Zhengcheng Zeng, Lingchang Kong, Yuanxin Chen and Pingli He *

State Key Laboratory of Animal Nutrition, College of Animal Science and Technology,
China Agricultural University, Beijing 100193, China
* Correspondence: hepingli@cau.edu.cn; Tel.: +86-10-62733688

Academic Editor: Jose M. Miranda
Received: 18 July 2019; Accepted: 30 August 2019; Published: 31 August 2019

Abstract: *N*-carbamylglutamate (NCG), a synthetic analogue of *N*-acetylglutamate, is an activator of blood ammonia conversion and endogenous arginine synthesis. Here, we established an accurate quantitative determination of NCG in feeds, animal tissues, and body fluids using the high performance liquid chromatography tandem mass spectrometry (HPLC-MS/MS). The sample pretreatment procedures included extraction with 0.5% of formic acid in water/methanol (80/20, *v/v*), and purification using an anionic solid phase extraction cartridge. Satisfactory separation of NCG was achieved in 20 min with the application of an Atlantis T3 column, and a confirmative detection of NCG was ensured by multiple reaction monitoring of positive ions. NCG spiked in feeds, tissues, and body fluids were evaluated in regard to linearity, sensitivity, recovery, and repeatability. Recoveries for different sample matrices were in the range of 88.12% to 110.21% with relative standard deviations (RSDs) less than 8.8%. Limits of quantification were within the range of 0.012 to 0.073 mg kg^{-1} and 0.047 to 0.077 μg mL^{-1} for solid and liquid samples, respectively. This study will provide a solid foundation for the evaluation of availability and metabolic mechanism of NCG in animals.

Keywords: *N*-carbamylglutamate; feeds; animal products; milk; HPLC-MS/MS

---

1. Introduction

N-carbamylglutamate (NCG) is a synthetic analogue of *N*-acetylglutamate (NAG) (Figure 1), which is the allosteric stimulator of carbamyl phosphate synthase-1 (CPS1). CPS1 is a key enzyme functioning in the urea cycle and endogenous arginine synthesis pathway [1]. Like NAG, NCG can activate CPS1 and lead the conversion of blood ammonia into the mitochondrial carbamoyl-phosphate, further stimulating the endogenous synthesis of arginine. Due to its good stability and safety [2,3], NCG was initially used for the clinical treating of hyperammonemia caused by the NAG synthetase deficiency [4,5], propionic aciduria and methylmalonic aciduria [6,7], and maple syrup urine disease [8]. NCG is also proven to be a novel, effective and low-cost substitute feed additive for arginine. In 2014, China's ministry of agriculture approved NCG as a new feed additive to use in the livestock (new feed additive certificate no. 2014-01). The maximum addition limit of NCG in the compound feed is 800 g t^{-1} of sow, and 880 g t^{-1} of dairy cow, respectively. Compared with the sole supplementation of NAG or arginine, NCG can avoid being catabolized by the deacylase or amino acid metabolism enzymes, and does not cause nutritional antagonism against other amino acids, especially lysine, tryptophan, and histidine [1,9]. Previous studies have indicated that oral administration of NCG 50 mg per kg body weight two times a day, increases piglets' plasma arginine concentration by 68%, and weight gain by 61% in 10 days [1]. More recently, several results have been reported that gilts with dietary supplementation of 0.05% and 0.1% of NCG had more pigs born alive, and significantly increased the live litter weight [10,11]. A similar effect of NCG on growth promotion and reproductive performance enhancement have also been found in other species of animals including the rat [12], cattle [13], sheep [14], goat [15], and yellow-feather broiler [16].

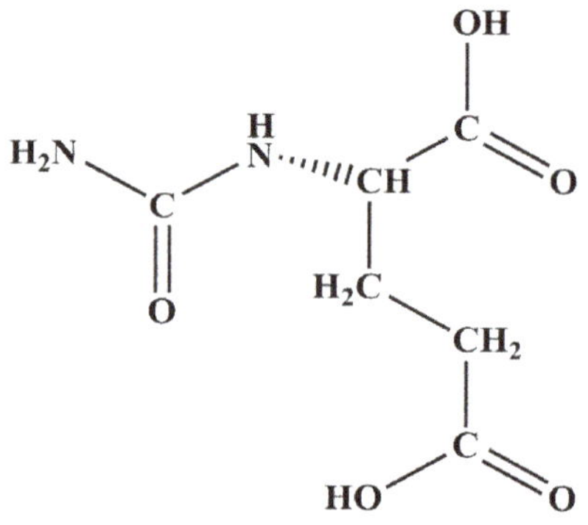

Figure 1. The chemical structure of *N*-carbamylglutamate.

Although the physiological functions of NCG have been well elucidated, and NCG has long been shown to be non-toxic and free of side effects [2], its metabolic process in animals remains unclear. With the promising use of NCG in the livestock, it is necessary to develop an accurate measurement method for NCG in the feed and animal products. Based on the chemical structure of NCG, the infrared spectroscopy (IR), nuclear magnetic resonance (NMR) spectroscopy, and high performance liquid chromatography (HPLC) have been adopted in the analysis of the structure and content of the NCG additive (purity > 97%) [17]. However, due to the complexity of the feed and animal products matrices, few studies have been conducted to deal with the determination of NCG in the feeds [18], and to the best of our knowledge, there is no former report regarding the NCG residue detection in animal tissues or body fluids.

Since NCG has poor absorption of the UV-Vis spectrum, using the HPLC based on the UV detection will result in low responsivity and sensitivity. Our group tried to determine the NCG using the before- or past-column derivation, but it showed a multiple matrix interference leading to unsatisfactory resolution results, and the procedures were relatively complex. In recent years, the high performance liquid chromatography-tandem mass spectrometry (HPLC-MS/MS), due to its high selectivity and sensitivity, has become the preferred method to analyze the trace small molecular compounds in the complex substrates [19,20]. This study developed a HPLC-MS/MS method to qualitatively and quantitatively determine NCG in the feed, animal tissues, and body fluids with excellent accuracy and sensitivity. The proposed method can achieve a fast separation of NCG within a 20 min gradient elution, and the method validation was achieved by evaluating the linearity, sensitivity, recovery, as well as the accuracy and repeatability for NCG in the feeds and animal products.

2. Results and Discussion

2.1. Optimization of HPLC-MS/MS Conditions

In order to improve the analytical sensitivity and selectivity, parameters of the mass spectrometry such as the ionization mode, source temperature, capillary voltage, nebulizer pressure, sheath gas flow, collision energy, and fragmentor voltage were optimized using the NCG standard solution. The result showed that the most abundant precursor ion of NCG was the pseudo-molecular ion $[M + H]^+$ at m/z 191.0 (Figure 2A and Figure S1). The MS/MS spectrum of this precursor ion is shown in Figure 2B. The most abundant product ion (m/z 84.0) was used for quantitation while the two relatively abundant product ions (m/z 130.0, m/z 148.0) were selected for qualification (Figure 2C–E). With the aim to increase the signal responses of characteristic product ions and their stability, the parameters (fragmentor, collision energy) in the second mass analyzer were optimized. The optimal fragmentor of the precursor ion (m/z 191.0) is 50 V. The optimal collision energy of m/z 84.0, m/z 130.0, and m/z 148.0 are 15 V, 10 V, and 5 V, respectively.

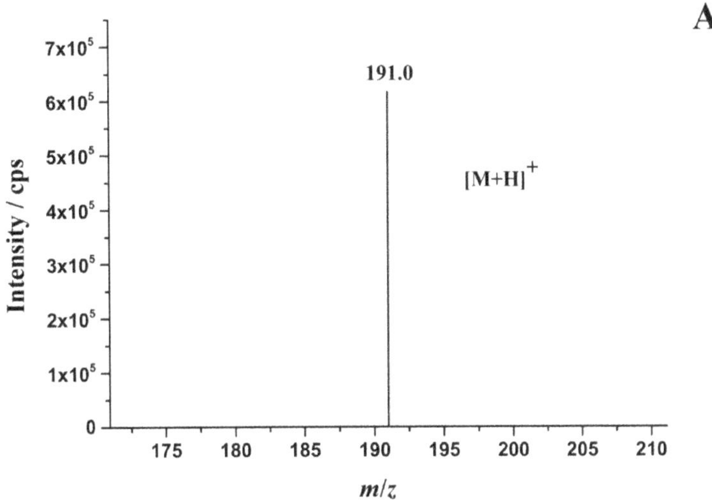

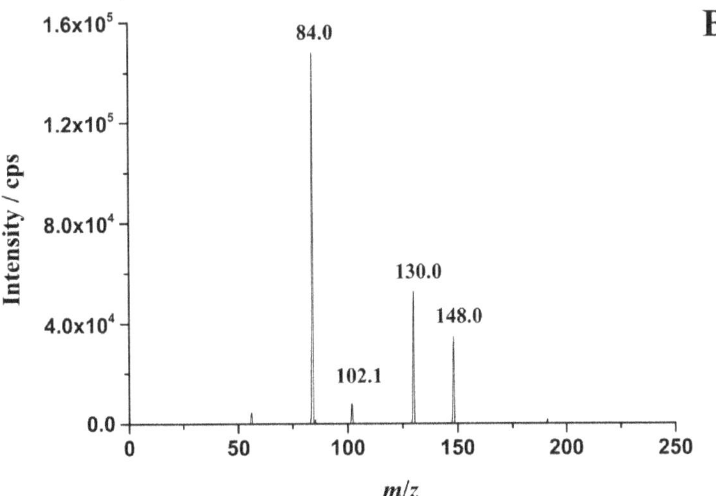

Figure 2. *Cont.*

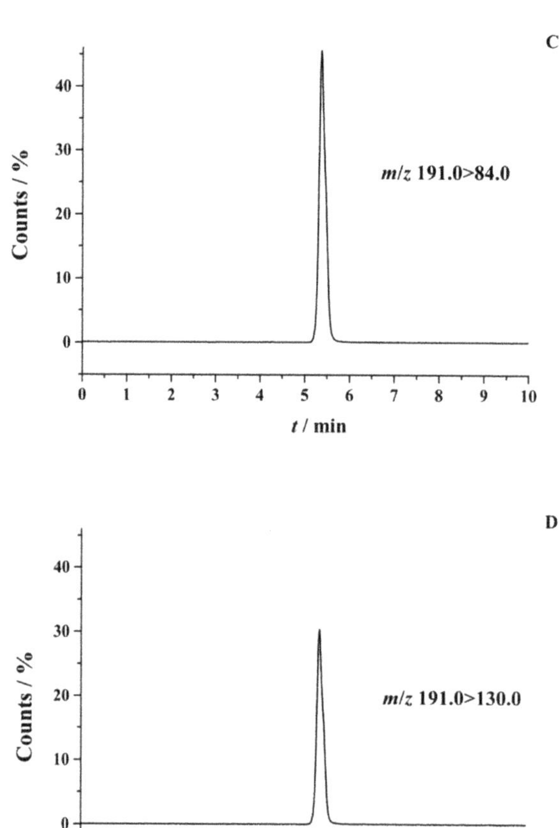

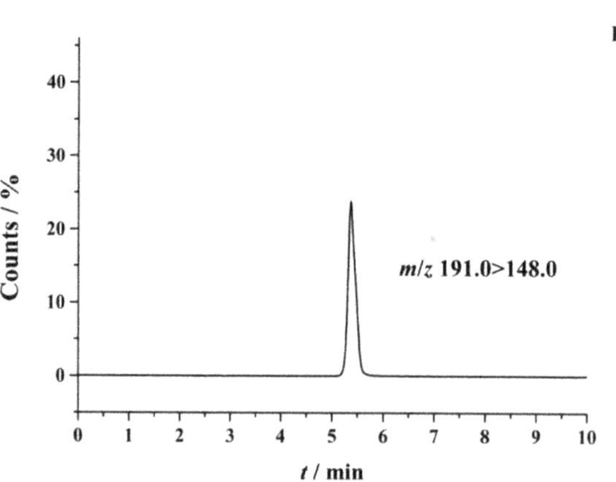

Figure 2. (**A**) Parent ion chromatogram of *N*-carbamylglutamate (NCG); (**B**) the tandem mass spectrometry (MS/MS) spectrum of the NCG precursor ion; (**C–E**) typical extracted mutiple reaction monitoring (MRM) ion chromatograms of the NCG standard conducted under optimal conditions.

NCG is a compound with strong polarity. A Waters Atlantis T_3 C_{18} column (4.6 × 250 mm, 5 μm) was compared with a Waters BEH C_{18} column (2.1 × 100 mm, 1.7 μm). The T_3 column showed a better retention for polar compounds with reduced peak tailing thus better peak shapes and improved analytical results than the BEH column. In addition, considering that there are many free amino acids, fatty acids, and other small molecular metabolites in animal samples, the long analytical column with strong polarity is more suitable for the separation of NCG without a precolumnar derivation. Thus, the Atlantis T3 C18 column was chosen as the analysis column. Moreover, the mobile phase of HPLC is a critical factor influencing the analytical results and sensitivity of the detection. Compared with acetonitrile, using methanol as the organic mobile solvent can achieve better chromatographic separation of NCG. Due to the weak acidity of NCG, the addition of a small amount of formic acid could improve the peak shape and improve the sensitivity. Therefore, 0.1% of the formic acid aqueous solution and methanol were selected to be the mobile phases for the binary gradient pump system. Figure 2C showed the typical extracted MRM ion chromatogram of NCG conducted under optimal conditions.

2.2. Optimization of Extract Condition

An effective pretreatment should serve the purpose that the impurity substance is excluded while the target analyte stays as much as possible. As NCG is soluble in both water and methanol, pilot experiments have been conducted to determine the optimal extraction solvent with the finest extraction result. For the feed samples, five extract solvents including water, water/methanol (80:20, *v/v*), water/methanol (50:50, *v/v*), water/methanol (20:80, *v/v*) and methanol were compared. The results show that the recoveries were water > water/methanol (80:20, *v/v*) > methanol > water/methanol (50:50, *v/v*) > water/methanol (20:80, *v/v*). However, using water as the extraction solvent was subjected to a greater matrix effect presumably due to the complex water-soluble additive in the feed. Thus, water/methanol (80:20, *v/v*) was selected as the extraction solvent. To better improve the extraction efficiency of NCG in the water/methanol (80/20, *v/v*) solvent, different concentration levels of formic acid in the water including 0, 0.1%, 0.5%, and 1% were compared. The results show that 0.5% of formic acid in water/methanol (80/20, *v/v*) manifested the best extraction result with a recovery of 99.1%. In addition, the high recovery (>100%) were shown in Figure 3C and D. We speculated that the matrix effect in the C and D extract may cause the enhancement of the NCG signal in MS, which leads to the recovery of NCG over 100%. Therefore, 0.5% of formic acid in water/methanol (80/20, *v/v*) was applied as the extraction solvent for NCG in the feeds (Figure 3). For the animal samples, the biggest interference came from the high lipid and protein content in the samples [21]. Similarly, different proportions of methanol and 0.5% of formic acid solution were also tested for the optimization of extraction agents. Moreover, the application of ice-cold methanol combined with the subsequent refrigerated centrifugation step can efficiently precipitate and remove most of the interference proteins present in milk, serum, and tissue samples [20,22]. Thus, ice-cold methanol (100%) was finalized as the optimal extraction solvent for NCG in the animal body fluids and tissues due to its precipitation effect of proteins and relatively good recovery (Figure 3).

2.3. Optimization of Purification Condition

A subsequent purification experiment of the sample extracts was conducted using the solid-phase extraction (SPE) cartridges to further decrease the matrix interference and enhance sensitivity. NCG is a weak acid compound. A strong anion exchange SPE cartridge was used in virtue of its strong adsorption ability for the anionic compound. For the better retentivity of NCG, an appropriate amount of 5% ammonia solution was added to adjust the pH of the extract and make NCG anionic. The results show that the pH of the extract that ranged from 8 to 10 was optimal, the NCG retained above 95% on the SPE (Figure 4). Since the cartridge was a strong anion exchange column, considering the isoelectric point of the NCG (pI 3.02), a lower pH (<8) will not be able to sufficiently make the NCG anionic, resulting in lower retentivity. On the other hand, if the pH is too high (>10), other interfering compounds may be turned anionic and compete with NCG for the limited retention capacity of the cartridge, this will also cause lower retentivity of NCG in the column. In addition, the NCG content in the feeds were very high, a direct application of the sample extracts may cause oversaturation and lead to unsatisfactory recoveries of NCG due to limited adsorption capacity of the cartridges. Our study has indicated that a reduction of the NCG concentration in the sample extracts could significantly increase the NCG recoveries. To avoid oversaturation and improve the peak shape, feed sample extracts were diluted 10 times with water before loaded onto the cartridges. On the contrary, if the content of the NCG in the sample is too low, such as the biological sample, the loading volume should be increased before passing through the column.

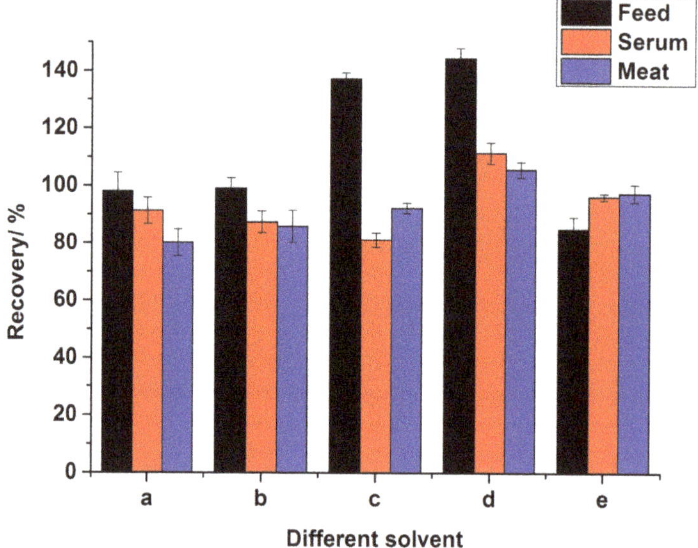

Figure 3. Recoveries of NCG from matrices using different solvents: (a) 0.5% of formic acid solution; (b) 0.5% of formic acid solution/methanol (80:20, *v/v*); (c) 0.5% of formic acid solution/methanol (50:50, *v/v*); (d) 0.5% of formic acid solution/methanol (20:80, *v/v*); (e) methanol.

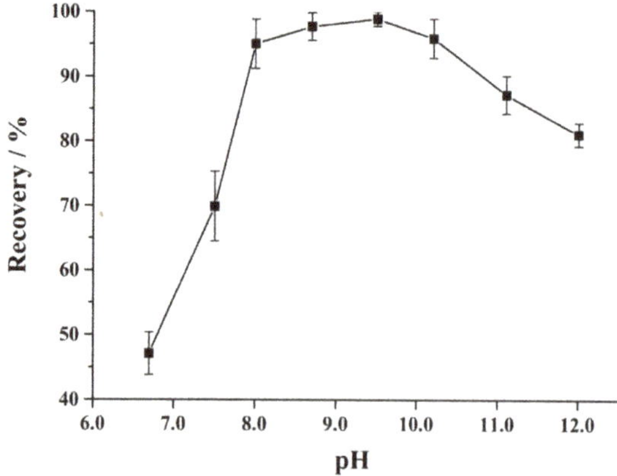

Figure 4. Dependence of the retention abilities of the NCG on the solid-phase extraction (SPE) cartridge with different pH of the loading solvent.

2.4. Method Validation

2.4.1. Stability and Matrix Effect

For the stability test, the NCG stock solution was diluted into 0.005, 0.1, 2 μg mL^{-1} to have six aliquots with concentrations around the linearity range. These aliquots were stored at −4 °C and 20 °C, respectively. Normally, the criterion for the stability is that degradation is equal to or less than 5%. The results show that the solution could be stored stably for at least three months and one week at the −4 °C refrigerator and 20 °C, respectively.

Matrix effects come from various chemical and physical processes during ionization of the analytes in the ESI mode of mass spectrometry, which causes signal suppression or enhancement. The matrix effect was evaluated by comparing three NCG concentrations (0.005, 0.1, 2 μg mL^{-1}) constructed in the solvent and fortified sample extract. The matrix effects were calculated by comparing the peak area of NCG in the solvent with the peak area of NCG in the fortified sample extract. The results show that the matrix effects ranged from 1.08 to 0.54 for the feed and biological samples. Especially for the biological samples, the matrix inhibition effects were obvious when a high concentration NCG was added. Therefore, the quantification of NCG must be performed using the matrix-matched calibrators.

2.4.2. Linearity, Precision, Accuracy, LOQ and LOD

A series of experiments were carried out with the aim to validate the linearity, sensitivity, precision, and accuracy of the established method. The NCG stock solution was diluted into 0.0016, 0.08, 0.2, 0.5, 1, 2 μg mL^{-1} working solutions. Excellent linearity was obtained at the NCG concentrations ranging from 0.0016–2 μg mL^{-1} with a correlation coefficient of (R^2) 0.999. To evaluate the sensitivity of this method, the limits of detection (LOD) and the limits of quantitation (LOQ) were tested. The LOD, defined as the analyte concentration at three times the signal to noise ratio (S/N), was determined for NCG in the feed, milk, serum, meat, liver, and kidney samples. The results indicated that the LOD values were within the range of 0.0035 to 0.022 mg kg^{-1} and 0.014 to 0.023 μg mL^{-1} for the solid and liquid samples, respectively (Tables 1 and 2). The LOQ, defined as the analyte concentration at 10 times the S/N, were in the range of 0.012 to 0.073 mg kg^{-1} and 0.047 to 0.077 μg mL^{-1} for the solid and liquid samples, respectively (Tables 1 and 2). Here, the compound feed, concentrated feed, and premixed feed are all called the feed. Since the three matrices are similar, the LOD and LOQ values for the three feeds were just slightly different. For convenient use, we have integrated the three similar results into one LOD and one LOQ.

Table 1. Percentage recoveries and relative standard deviations (RSDs) of NCG in different feed matrices (n = 3).

Matrices	Added Concentration (%)	Recoveries (%)	LOD (mg kg^{-1})	LOQ (mg kg^{-1})
Compound feed	0.01	96.85 (1.3) [a]		
	0.05	98.89 (2.9)		
	0.1	99.24 (5.0)		
Concentrated feed	0.05	107.0 (3.5)		
	0.2	106.1 (7.1)	0.022	0.073
	1.0	97.60 (8.8)		
Premix	0.5	94.27 (6.8)		
	2.0	91.90 (3.9)		
	10	100.2 (3.6)		

[a] Figures in bracket represented the relative standard deviation (%).

Table 2. Percentage recoveries and relative standard deviations (RSDs) of NCG in different animal samples (n = 3).

Matrices	Added Concentration	Recoveries (%)	LOD (μg mL^{-1})	LOQ (μg mL^{-1})
Serum (μg mL^{-1})	0.1	94.25(3.8) [a]	0.023	0.077
	1.0	110.2 (2.3)		
	10	96.34 (5.1)		
Milk (μg mL^{-1})	0.1	88.12 (1.8)	0.014	0.047
	1.0	102.5 (3.2)		
	10	89.58 (4.7)		
Meat (mg kg^{-1})	0.05	93.84 (7.6)	0.0038	0.013
	0.1	101.6 (3.4)		
Kidney (mg kg^{-1})	0.05	89.14 (3.1)	0.0069	0.023
	0.1	93.65 (5.6)		
Liver (mg kg^{-1})	0.05	94.54 (7.4)	0.0035	0.012
	0.1	97.36 (2.9)		

[a] Figures in bracket represented the relative standard deviation (%).

The method precision and accuracy were evaluated by the recoveries of NCG in the spiked samples. Recoveries were determined for the feeds, milk, serum, meat, procine liver, and kidney, three replicates were used for each sample at different concentrations. The MRM ion chromatograms for a blank (free of NCG) feed sample and the compound feed sample spiked with NCG were demonstrated in Figure 5A,B and Figures S2–S19, respectively. The peak intensity ratio of the three product ions of NCG can be used to confirm the presence of NCG when compared with the peak intensity ratio obtained from the standard sample. The most abundant product ion of the mass spectrum was selected for the quantitative determination and evaluation of the recoveries for NCG. Calibration curves developed using an external standard were used to quantify NCG in the spiked samples. The spiked levels, spiked recoveries, and coefficients of variation were evaluated. According to the recommended dose of NCG in the feeds, NCG was spiked in the compound feed at 0.01%, 0.05%, and 0.1%, the premix feed at 0.05%, 0.2%, and 1%, and the concentrate feed at 0.5%, 2%, and 10%. Three replicates were analyzed for each concentration. A good consistency was found between the actual spiked amount of NCG and the estimated concentrations in three different feed matrices (Table 1). The recoveries were within the range of 91.90% to 107.0% with the coefficients of variation ranging from 1.3% to 8.8%.

Similarly, the above established method was also applied to the detection of the NCG spiked animal products. The MRM ion chromatograms for the blank and spiked animal products samples were shown in Figure 5C–G and Figures S20–S49, respectively. NCG was spiked in meat, kidney, and liver at 0.05, and 0.1 mg kg^{-1}. For each sample concentration, three replicates were analyzed to evaluate the spiked levels, spike recoveries, and coefficients of variation. As presented in Table 2, reliable results were found for most test samples. The recoveries for NCG ranged from 89.14% to 101.56% and the coefficients of variation were less than 7.6%. The NCG were spiked in milk and serum at 0.1, 1, and 10 μg mL^{-1}. The recoveries ranged from 88.12% to 110.21% with the coefficients of variation ranging from 1.8% to 5.1% (Table 2).

Both the intra-assay and inter-assay reproducibility of this method were evaluated. For instance, five repeats at an intermediate concentration (0.05%) were used to determine the recoveries of NCG spiked in the compound feed. According to the analytical results, the recoveries of NCG were within the range of 98.89% to 102.76% and the coefficients of variation were less than 2.9% and 3.8% for the intra-assay (within a day) and inter-assay (over a period of five consecutive days) measurements, respectively. For the animal samples, this method also showed satisfactory reproducibility. For 1 μg mL^{-1} of NCG spiked in the serum, the recoveries of NCG were within the range of 93.25% to 104.68% and the coefficients of variation were less than 4.7% and 5.4% for the intra-assay and inter-assay measurements, respectively. For 0.05 mg kg^{-1} of NCG spiked in the liver, the recoveries of NCG were within the range of 91.23% to 106.71% and the coefficients of variation were less than 3.9% and 6.7% for the intra-assay and inter-assay measurements, respectively.

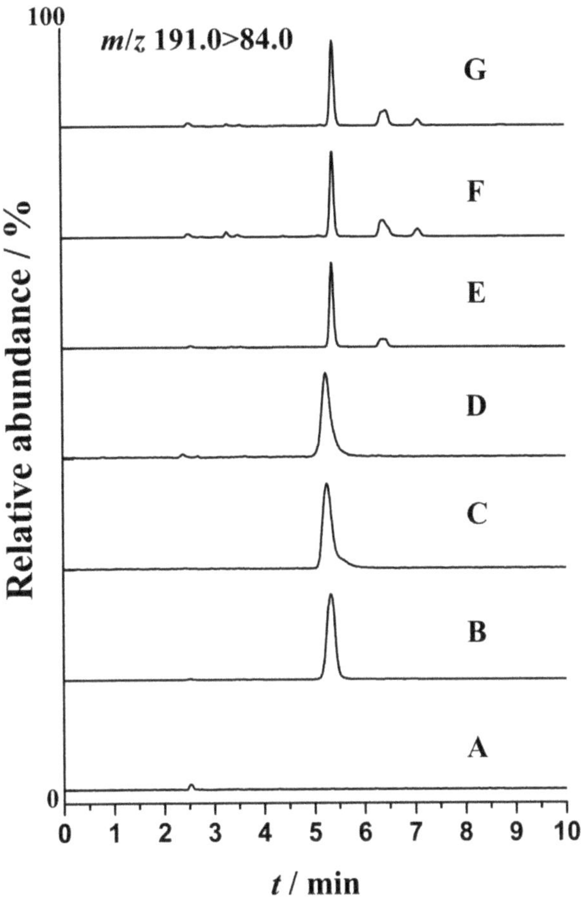

Figure 5. Quantitative product ion chromatograms of the NCG in different matrices: (**A**) Blank feed sample; (**B**) compound feed; (**C**) milk; (**D**) serum; (**E**) meat; (**F**) liver; (**G**) kidney.

2.5. The Analysis of Authentic Samples

The method established in this study was applied to determine the NCG content in the authentic feed samples added with NCG additive. The results showed that NCG was detected in the compound feed, concentrated feed, and premix feed samples with the average content of 0.098%, 0.95%, and 9.53%, respectively. The variation between the two parallel samples was less than 10%, while NCG was not detected in the ordinary swine feed samples (<0.022 mg kg^{-1}). According to the actual addition amount, the accuracy for the determination of the compound, concentrate, and premix feed was 98.0%, 95.0%, and 95.3%, respectively. The proposed method was also adopted for the determination of NCG in the authentic serum samples from pigs and milk sample from cows (each concentration has six replicates). The results showed that the concentration of NCG in the serum samples increased significantly with the increase of NCG in the diets (Table 3). Except for the control group, the NCG levels in the serum of four diets ranged from 0.11 to 1.31 µg mL^{-1}. Moreover, the multiple relationship between the serum NCG and dietary NCG was basically consistent. The NCG concentration in milk was about

0.0647 µg mL^{-1} when the cow was fed 20 g of NCG per day. Therefore, the developed method proved both feasible and reliable for the quantitative determination of NCG in the routine analysis.

Table 3. The results of NCG in the authentic samples (n = 6).

Sample	Diets Added (%)	NCG Concentration (µg mL^{-1})	Intra-Day Precision CV (%)	Inter-Day Precision CV (%)
Serum [b]	0	ND [a]	-	-
	0.025	0.112	6.8	8.7
	0.05	0.253	2.4	5.6
	0.1	0.561	7.8	11
	0.2	1.31	12	5.1
Milk	0	ND	-	-
	20 g/day	0.0647	11	7.4

[a] Not detected, [b] Serum from piglets.

3. Materials and Methods

3.1. Materials and Reagents

The NCG standard (purity >99.5%) was purchased from Sigma-Aldrich (St. Louis, MO, USA). The methanol and formic acid used were of a HPLC grade and were obtained from Fisher Scientific International (Hampton, NH, USA) and Dikma Technology (Richmond Hill, ON, Canada), respectively. Other chemicals used were all of the analytical grade. A Milli-Q system (Millipore Corporation, Bedford, MA, USA) was applied to provide ultrapure water for the preparation of all aqueous solutions.

The NCG stock solution (1 mg mL^{-1}) was prepared by dissolving the NCG standard in ultrapure water, and preserved at 4 °C. The preparation of the NCG working solutions was done daily by diluting the stock solution to appropriate concentrations with ultrapure water. The extraction solvent, 0.5% of formic acid in water/methanol (80/20, *v/v*), was prepared by mixing 4 mL of formic acid with 796 mL of Milli-Q water, and then mixing with 200 mL of methanol.

3.2. Instruments and Apparatus

The HPLC-MS/MS was performed on an Agilent 1200 UHPLC system coupled with an Agilent 6460 Triple Quadrupole Mass Spectrometer (Agilent Technologies, Fermont, CA, USA). An Ultrasonic Cleaner (Kunshan, China) was used to promote the sample dissolution and extraction. A low temperature high speed centrifuge (Eppendorf, Hamburg, German) was used to centrifuge the samples.

3.3. Sample Preparation

Feed samples free of NCG were provided by the Ministry of Agricultural Feed Industry Centre (Beijing, China). The liver and kidney as blanks were from a porcine, and milk was from a cow. They were all purchased from local supermarkets. The serum of the pig as blanks were collected from the animal research station of China Agricultural University (Hebei, China). Prior to the extraction, the feed samples were crushed into about 60 mesh by a small mill. The meat, liver, and kidney tissues were homogenized in a homogenizer for five min. The milk and serum were pre-treated by centrifuging at 14,000 rpm for 10 min at 4 °C, then the supernatant was collected.

3.4. Sample Extraction and Purification

Extraction of the feed samples was carried out by adding 20 mL of the extraction solvent (0.5% of formic acid in water/methanol, 80/20, *v/v*) into the feeds (2 g for the compound feed, 0.2 g for the concentrated feed and premix). Usually, the recommended amounts of NCG in the pig compound feed, premixed feed, and concentrated feed were about 0.05%, 1% and 10%, respectively. Since the concentration of NCG in the premix or concentrated feed was too high, in order to avoid too much dilution before the HPLC-MS/MS analysis, the sample amount in the premix or concentrated feed was relatively reduced. After the vortex mixing and a 30 min ultrasonic bath, 10 min of centrifugation at 14,000 rpm at 4 °C was completed and the supernatant was collected. For the fluid samples, 2 mL of pre-centrifuged milk or serum was pipetted into a Corning® 15 mL centrifuge tube, and 8 mL of an ice-cold extraction solvent was added. After the vortex mixing, the samples were centrifuged at 14,000 rpm for 10 min at 4 °C and the supernatant collection was done. For the tissue analyses, 2 g of tissues samples (meat, kidney, or liver) were extracted by adding 20 mL of an ice-cold extraction solvent. After a thorough vortex mixing, the samples were placed on ice for 30 min and centrifuged for 10 min at 14,000 rpm at 4 °C, then the supernatant collection was done.

Before further purification, the above supernatant was diluted 10 times using water and adjusted by adding an appropriate amount of 5% of ammonia solution to pH 8–10. The strong anion exchange cartridge (Agilent SampliQ SAX, 200 mg, 3 mL, Agilent Technologies, Fermont, CA, USA) was firstly preconditioned with 3 mL of methanol following by 3 mL of water. Then, 3 mL of the diluted supernatant was loaded and slowly passed through the cartridge with a flow rate of approximately 1 mL min^{-1}. After rinsing with 3 mL 5% of the ammonia solution and 3 mL of methanol, elution was carried out with 3 mL 2% of formic acid in methanol. The eluate was then collected and gently evaporated under a nitrogen stream in a 50 °C water bath. The residue was reconstituted with 1 mL 0.1% of an aqueous formic acid solution and filtered by a 0.1 µm Syringe Filter (Tianjin Fuji Science and Technology, China). 10 µL of the sample was injected by the autosampler for the HPLC-MS/MS analysis.

3.5. HPLC-MS/MS Conditions

A Waters Atlantis T3 column (4.6 × 250 mm, 5 µm) was applied in the chromatographic separation of NCG. The column temperature was maintained at 35 °C. The mobile phases consisted of solution A (0.1% of an aqueous formic acid solution) and B (methanol), and were delivered at a flow rate of 0.8 mL min^{-1}. The total run time was set at 20 min. A gradient elution program was developed as follows: 90% of solution A (initial), with 90–15% of solution A (from 6 to 6.1 min), 15–90% of solution A (from 12 to 12.1 min). Before the next injection, a 7.9 min equilibration was necessary. A calibration curve calculated from the NCG standard solution was used to quantify the NCG content in the spiked samples.

The electron spray ionization source was operated under an optimized condition: Capillary voltage 3500 V, nebulizing gas temperature 350 °C, nebulizing gas flow 5 L min^{-1}, sheath gas temperature 350 °C, and sheath gas flow 7 L min^{-1}. Positive ions were monitored. The multiple reaction monitor (MRM) mode was applied for the quantitation analysis. Data were collected and processed on the Agilent MassHunter Workstation software (Version B.04.00).

3.6. The Method Validation

The method validation was performed with the feeds, serum, meat, liver, and kidney samples. Extracts of the NCG-free samples were used for the preparation of matrix-matched calibration standards. The artificially prepared samples were spiked with the NCG standard solution of 0.01–10% in the feeds, 0.1–10 µg mL^{-1} in fluids, or 0.05–0.1mg kg^{-1} in tissues. The validity of the method including the stability, matrix effect, linearity, sensitivity, as well as the precision and accuracy of the method were evaluated in three replicates.

3.7. The Analysis of Authentic Samples

One hundred kg of the commercially prepared compound feed, concentrated feed and premix were added with the NCG feed additive product (NCG content ≥ 97%) obtained from Ya Tai XingMu company (Beijing, China) to the concentrations of 0.1% for the compound feed, 1% for the concentrated feed and 10% for the premix, respectively.

The animals breeding experiments were carried out at the animal research station of China Agricultural University (Hebei, China). All piglets used in this study were housed and handled according to the established guidelines of China Agricultural University, which were approved by the China Agricultural University Animal Care and Use Committee. Thirty weaned piglets (each 30–40 kg in weight) were randomly assigned to one of the five groups (n = 6 piglets per group). Five pig diets were prepared with NCG added 0%, 0.025%, 0.05%, 0.1%, and 0.2%, respectively. After 26 fed days, blood samples from the precaval veins were collected after fasting 12 h and centrifuged (3000 rpm) for 10 min. Then, the supernatant was transferred into new tubes and stored at −20 °C until use. Milk samples were collected from the cow before and after feeding 20 g of NCG per day.

Thirty weaned piglets (each 30–40 kg in weight) were randomly assigned to one of the five groups (n = 6 piglets per group).

4. Conclusions

A fast, sensitive and reliable HPLC-MS/MS method was successfully developed and optimized for the qualitative and quantitative determination of NCG in the feeds, animal tissues, and body fluids. Based on the high selectivity and satisfactory separation result of the HPLC-MS/MS technique, this method can achieve confirmative identification and accurate determination of NCG in the complex matrices including feeds and animal products. Furthermore, the applicability of the established method has been validated using the real feed, serum and milk samples.

Supplementary Materials: The following are available online at http://www.mdpi.com/1420-3049/24/17/3172/s1.

Author Contributions: Y.M. and Z.Z. designed and performed the experiments; L.K. and Y.C. analyzed the data; P.H. and Y.M. wrote the paper.

Funding: The financial support from The National Key Research and Development Program of China (2018YFD0500402) is gratefully acknowledged.

Conflicts of Interest: The authors declare no conflict of interest.

References

1. Wu, G.; Knabe, D.A.; Kim, S.W. Arginine nutrition in neonatal pigs. *J. Nutr.* **2004**, *134*. [CrossRef] [PubMed]
2. Harper, M.S.; Shen, Z.A.; Barnett, J.F.; Krsmanovic, L.; Myhre, A.; Delaney, B. N-acetyl-glutamic acid: Evaluation of acute and 28-day repeated dose oral toxicity and genotoxicity. *Food Chem. Toxicol.* **2009**, *47*, 2723–2729. [CrossRef] [PubMed]
3. Wu, X.; Wan, D.; Xie, C.; Li, T.; Huang, R.; Shu, X.; Ruan, Z.; Deng, Z.; Yin, Y. Acute and sub-acute oral toxicological evaluations and mutagenicity of N-carbamylglutamate (NCG). *Regul. Toxicol. Pharmacol.* **2015**, *73*, 296–302. [CrossRef] [PubMed]
4. Schubiger, G.; Bachmann, C.; Barben, P.; Colombo, J.P.; Tönz, O.; Schüpbach, D. N-Acetylglutamate synthetase deficiency: Diagnosis, management and follow-up of a rare disorder of ammonia detoxication. *Eur. J. Pediatr.* **1991**, *150*, 353–356. [CrossRef] [PubMed]
5. Burlina, A.B.; Bachmann, C.; Wermuth, B.; Bordugo, A.; Ferrari, V.; Colombo, J.P.; Zacchello, F. Partial N-acetylglutamate synthetase deficiency: A new case with uncontrollable movement disorders. *J. Inherit. Metab. Dis.* **1992**, *15*, 395–398. [CrossRef]
6. Gebhardt, B.; Vlaho, S.; Fischer, D.; Sewell, A.; Böhles, H. N-carbamylglutamate enhances ammonia detoxification in a patient with decompensated methylmalonic aciduria. *Mol. Genet. Metab.* **2003**, *79*, 303–304. [CrossRef]
7. Gebhardt, B.; Dittrich, S.; Parbel, S.; Vlaho, S.; Matsika, O.; Bohles, H. N-Carbamylglutamate protects patients with decompensated propionic aciduria from hyperammonaemia. *J. Inherit. Metab. Dis.* **2005**, *28*, 241–244. [CrossRef]
8. Kalkan, U.S.; Coker, M.; Habif, S.; Ulas, E.S.; Karapinar, B.; Ucar, H.; Kitis, O.; Duran, M. The first use of N-carbamylglutamate in a patient with decompensated maple syrup urine disease. *Metab. Brain. Dis.* **2009**, *28*, 241–244. [CrossRef]
9. Wu, G.; Bazer, F.W.; Davis, T.A.; Jaeger, L.A.; Johnson, G.A.; Kim, S.W.; Knabe, D.A.; Meininger, C.J.; Spencer, T.E.; Yin, Y.L. Important roles for the arginine family of amino acids in swine nutrition and production. *Livest. Sci.* **2007**, *112*, 8–22. [CrossRef]

10. Liu, X.; Wu, X.; Yin, Y.; Liu, Y.; Geng, M.; Yang, H.; Blachier, F.; Wu, G. Effects of dietary l-arginine or N-carbamylglutamate supplementation during late gestation of sows on the miR-15b/16, miR-221/222, VEGFA and eNOS expression in umbilical vein. *Amino Acids.* **2012**, *42*, 2111–2119. [CrossRef]

11. Zhang, B.; Che, L.; Lin, Y.; Zhuo, Y.; Fang, Z.; Xu, S.; Song, J.; Wang, Y.; Liu, Y.; Wang, P.; et al. Effect of dietary N-carbamylglutamate levels on reproductive performance of gilts. *Reprod Domest. Anim.* **2014**, *49*, 740–745. [CrossRef] [PubMed]

12. Liu, G.; Xiao, L.; Cao, W.; Fang, T.; Jia, G.; Chen, X.; Zhao, H.; Wu, C.; Wang, J. Changes in the metabolome of rats after exposure to arginine and N-carbamylglutamate in combination with diquat, a compound that causes oxidative stress, assessed by 1H NMR spectroscopy. *Food Funct.* **2016**, *7*, 964–974. [CrossRef] [PubMed]

13. Feng, T.; Schütz, L.; Morrell, B.; Perego, M.C.; Spicer, L. Effects of N-carbamylglutamate and L-arginine on steroidogenesis and gene expression in bovine granulosa cells. *Anim. Reprod. Sci.* **2018**, *188*, 85–92. [CrossRef] [PubMed]

14. Zhang, H.; Sun, L.; Wang, Z.; Deng, M.; Nie, H.; Zhang, G.; Ma, T.; Wang, F. N-carbamylglutamate and L-arginine improved maternal and placental development in underfed ewes. *Reproduction* **2016**, *151*, 623–635. [CrossRef] [PubMed]

15. Wang, Z.; Wang, R.; Meng, C.; Ji, Y.; Sun, L.; Nie, H.; Mao, D.; Wang, F. Effects of dietary supplementation of N-Carbamylglutamate on lactation performance of lactating goats and growth performance of their suckling kidlets. *Small Rumin. Res.* **2019**, *175*, 142–148. [CrossRef]

16. Hu, Y.; Shao, D.; Wang, Q.; Xiao, Y.; Zhao, X.; Shen, Y.; Zhang, S.; Tong, H.; Shi, S. Effects of dietary N-carbamylglutamate supplementation on growth performance, tissue development and blood parameters of yellow-feather broilers. *Poult. Sci.* **2019**, *98*, 2241–2249. [CrossRef] [PubMed]

17. Shu, X.; Wu, C.-L.; Wan, K.; Zhang, M.; Wu, X. Synthesis, Crystal Structure and Spectroscopic Properties of Zn(II) with N-carbamylglutamate Ligand. *Chin. J. Struc. Chem.* **2015**, *34*, 423–427.

18. Tang, S.; Wang, Y.; Wen, P.; Xin, Z. Determination of N-carbamyl-L-glutamic acid in feedstuff by high performance liquid chromatography coupled with electrospray ionization tandem mass spectrometry. *Chinese J. Chromatogr.* **2014**, *32*, 184–188. [CrossRef]

19. Liu, R.; Hei, W.; He, P.; Li, Z. Simultaneous determination of fifteen illegal dyes in animal feeds and poultry products by ultra-high performance liquid chromatography tandem mass spectrometry. *J. Chromatogr. B* **2011**, *879*, 2416–2422. [CrossRef]

20. Yin, B.; Li, T.; Zhang, S.; Li, Z.; He, P. Sensitive analysis of 33 free amino acids in serum, milk, and muscle by ultra-high performance liquid chromatography-quadrupole-qrbitrap high resolution mass spectrometry. *Food Anal. Methods* **2016**, *9*, 2814–2823. [CrossRef]

21. Dhondt, L.; Croubels, S.; De Cock, P.; De Paepe, P.; De Baere, S.; Devreese, M. Development and validation of an ultra-high performance liquid chromatography–tandem mass spectrometry method for the simultaneous determination of iohexol, p-aminohippuric acid and creatinine in porcine and broiler chicken plasma. *J. Chromatogr. B* **2019**, *1117*, 77–85. [CrossRef] [PubMed]

22. De Baere, S.; Devreese, M.; Watteyn, A.; Wyns, H.; Plessers, E.; De Backer, P.; Croubels, S. Development and validation of a liquid chromatography–tandem mass spectrometry method for the quantitative determination of gamithromycin in animal plasma, lung tissue and pulmonary epithelial lining fluid. *J. Chromatogr. A.* **2015**, *1398*, 73–82. [CrossRef] [PubMed]

Sample Availability: Samples of the compounds are not available from the authors.

Article

Sequence Identification of Bioactive Peptides from Amaranth Seed Proteins (*Amaranthus hypochondriacus* spp.)

Alexis Ayala-Niño [1], Gabriela Mariana Rodríguez-Serrano [2],
Luis Guillermo González-Olivares [1,*], Elizabeth Contreras-López [1], Patricia Regal-López [3] and
Alberto Cepeda-Saez [3]

[1]   Chemistry Investigation Center, Universidad Autónoma del Estado de Hidalgo,
      Carretera Pachuca-Tulancingo km 4.5, Mineral de la Reforma Hidalgo C.P. 46067, Mexico
[2]   Biotechnology Department, Universidad Autónoma Metropolitana, Unidad Iztapalapa,
      Mexico City C.P. 55355, Mexico
[3]   Universidad de Santiago de Compostela, Campus Lugo, 15705 Santiago de Compostela,
      27002 A Coruña, Spain
*   Correspondence: lgonzalez@uaeh.edu.mx; Tel.: +52-771-717-2000

Received: 1 June 2019; Accepted: 3 August 2019; Published: 21 August 2019

Abstract: *Amaranthus hypochondriacus* spp. is a commonly grown cereal in Latin America, known for its high protein content. The objective of this study was to separate and identify bioactive peptides found in amaranth seeds through enzymatically-assisted hydrolysis using alcalase and flavourzyme. Hydrolysis was carried out for each enzyme separately and compared to two-step continuous process where both enzymes were combined. The biological activity of the resulting three hydrolysates was analyzed, finding, in general, higher bioactive potential of the hydrolysate obtained in a continuous process (combined enzymes). Its fractions were separated by RP-HPLC, and their bioactivity was analyzed. In particular, two fractions showed the highest biological activity as ACE inhibitors with IC50 at 0.158 and 0.134, thrombin inhibitors with IC50 of 167 and 155, and antioxidants in ABTS assay with SC50 at 1.375 and 0.992 mg/L, respectively. Further sequence analysis of the bioactive peptides was carried out using MALDI-TOF, which identified amino acid chains that have not been reported as bioactive so far. Bibliographic survey allowed identification of similarities between peptides reported in amaranth and other proteins. In conclusion, amaranth proteins are a potential source of peptides with multifunctional activity.

Keywords: amaranth protein; flavourzyme; alcalase; bioactive peptides; hydrolysates

---

1. Introduction

Many diseases that prevail nowadays could be tackled more efficiently, and even prevented, by combining a healthy diet with functional foods intake [1]. Among functional food components, we can find bioactive peptides, which are generally short sequences of amino acids encrypted in food proteins [2]. Bioactive peptides are released from proteins as a result of microorganisms' metabolic activity during fermentation, by proteolytic enzymes, or finally by the action of gastrointestinal enzymes once proteins are ingested. Application of commercially available enzymes has become a simple and inexpensive way to access free amino acids and peptides from proteins [3].

Consequently, a vast number of studies using various proteinases target the preparation of hydrolysates from different protein sources, such as cereals and pseudocereals, in an ongoing effort to obtain highly active biopeptides [4–7]. One of the most important highly-consumed Mesoamerican original seeds, with a high content of proteins and an excellent amino acid balance, is amaranth.

In Mexico, amaranth is consumed in a fresh form, offered by traditional confectionary and in typical dishes of Mexican cuisine [8]. Bioactive peptides found in amaranth exhibit various biological activities—such as anticholesterolemic, antihypertensive, antioxidant, and antithrombotic—and they are released mainly by in vitro digestion [9–12]. However, multiple bioactivities shown by the same peptide fraction of amaranth have never been studied.

In order to find peptide sequences with different biological activities, our research work focused on the separation and identification of the peptide sequences from amaranth proteins released during hydrolysis with two commercial enzymes, alcalase and flavourzyme, in separate and continuous hydrolysis processes. The multiple biological activities exhibited by peptides released from amaranth proteins through enzymatically-assisted hydrolysis have not been reported until now.

2. Results

2.1. Free Amine Groups Analysis during Enzymatic Hydrolysis

Once the hydrolitic enzymes were added to the protein sample, the progress of the reaction was monitored every 20 min by the measurement of free amine groups concentration released during the reaction course (Figure 1). It was observed that after 120 min of hydrolysis with alcalase (H1), the concentration of free amine groups equaled to 6170.53 ± 29.5 mg/L NH:⁻, whereas flavourzyme hydrolysis (H2) reached the highest concentration of free amine groups (5551.11 ± 33.83 mg/L NH:⁻) after 90 min of reaction. It was concluded that, in terms of enzymatic efficiency, alcasase overdoes flavourzyme in the release of peptide fractions. Following the analysis of H1 and H2, the combined two-step hydrolysis (H3) was carried out, and a final free amine groups concentration of 7468.89 ± 34.79 mg/L NH:⁻ was obtained after 40 min of reaction. This result proves that the continuous hydrolitic process (H3) was more efficient than using alcalase and flavourzyme separately.

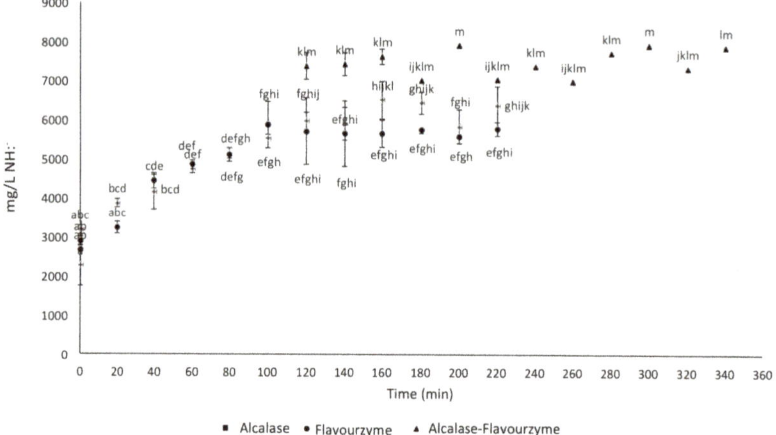

Figure 1. Free amine groups concentration during amaranth protein hydrolysis. * Different letters indicate statistically significant differences between treatments.

2.2. Hydrolysates Bioactivity Analysis

Table 1 shows the assessment of the different biological activities exhibited by the hydrolysates H1, H2, and H3, determined during amaranth proteins hydrolysis. For thrombin inhibition essay, H1 hydrolysate showed the highest inhibitory potential (90%), followed by H2 hydrolysate which inhibited thrombin activity by approximately 80%. No significant difference was observed for H3 and H1 hydrolysates as thrombin inhibitors. When angiotensin-converting enzyme inhibition was evaluated, it was observed that H3 hydrolysate showed the highest inhibitory potential (58%), in comparison to H1 and H2 hydrolysates (49% and 39%, respectively). Finally, antioxidant activity was measured by three different methods based on the free radical scavenging principle, such as ABTS, DPPH, and FRAP. The highest antioxidant activity was observed by the H3 sample in DPPH and FRAP assays (388.94 µmol TE/100 g and 592.54 µmol Fe2 E/100 g, respectively), while H1 hydrolysate exhibited the highest antioxidant activity measured by the ABTS method (425.86 ± 0.66 mg TE/100 g). Noteworthy, the antioxidant potential of H3 hydrolysate increased fivefold in the DPPH assay and ninefold in the FRAP assay, compared to the control sample of unhydrolyzed amaranth proteins.

Table 1. Bioactivity potential of amaranth protein hydrolysates after treatment with alcalase (H1), flavourzyme (H2), and in two-step combined hydrolysis (H3).

Hydrolysis	ACE Inhibition (%)	Thrombin Inhibition (%)	Antioxidant Activity		
			DPPH (µmol Trolox E/100 g)	ABTS (mg Trolox E/100 g)	FRAP (µmol Fe2 E/100 g)
Amaranth Protein	10.58 ± 1.19 [d]	11.90 ± 10.10 [c]	76.66 ± 1.60 [d]	115.65 ± 10.30 [d]	63.37 ± 5.72 [c]
H1	49.49 ± 1.47 [b]	92.85 ± 3.36 [a]	340.17 ± 10.95 [b]	425.86 ± 0.66 [a]	241.70 ± 9.38 [b]
H2	39.77 ± 2.15 [c]	80.95 ± 13.46 [b]	274.03 ± 10.84 [c]	398.36 ± 3.62 [c]	226.29 ± 11.20 [b]
H3	58.53 ± 2.58 [a]	92.85 ± 3.36 [a]	388.94 ± 2.73 [a]	404.90 ± 1.52 [b]	592.54 ± 29.29 [a]

Values are shown as mean ± standard deviation ($n = 3$); values in the same column with different superscript letters are significantly different ($p < 0.05$).

2.3. RP-HPLC Separation and Fraction Bioactivity

Since the H3 hydrolysate (obtained by application of continuous enzymatic hydrolysis) showed the highest bioactivity, it was chosen for further analysis in order to separate fractions and identify peptide sequences. In total, 56 fractions were obtained through RP-HPLC separation (see Appendix A), of which only 14 could be identified with known proteins. Moreover, protein content was determined in samples collected after 27 min (Table 2). In the case of the 14 fractions identified with protein, their bioactivity was evaluated by measurement of the IC50 in the ACE and thrombin inhibition assays, and SC50 in the ABTS assay. Results are gathered in Table 2.

Table 2. Bioactivities of amaranth protein fractions (Angiotensin I-Converting Enzyme and Thrombin inhibitory activity (IC$_{50}$, mg/L)) and antioxidant activity (ABTS radical scavenging; SC$_{50}$, mg/L).

Fraction	ACE (IC$_{50}$)	Thrombin (IC$_{50}$)	ABTS (SC$_{50}$)	Peptide Concentration (mg/L)
2	0.332 [cd]	38.46 [i]	4.204 [e]	0.2125
3	0.442 [e]	4.36 [h]	NI	0.8375
9	NI	0.426 [f]	NI	0.9062
18	0.614 [f]	2.65 [g]	2.538 [d]	0.7125
19	0.173 [b]	0.183 [b]	NI	0.3750
22	0.158 [ab]	0.167 [ab]	1.375 [b]	0.4687
23	NI	0.349 [e]	2.809 [d]	0.5625
27	0.808 [c]	0.402 [f]	1.616 [c]	0.3125
28	0.346 [d]	0.135 [a]	1.728 [c]	0.0937
32	0.192 [b]	0.298 [d]	6.931 [g]	0.4687
34	0.317 [cd]	0.247 [c]	2.593 [d]	0.4937
39	NI	0.247 [c]	5.561 [f]	0.4375
40	0.298 [c]	0.26 [cd]	4.547 [e]	0.8375
45	0.134 [a]	0.155 [a]	0.992 [a]	0.8125

Values for ACE, thrombin, and ABTS (mg peptide/mL) are mean ± SD ($n = 3$); values in the same column with different superscript letters are significantly different ($p < 0.05$). NI: not identified.

For the antihypertensive potential, the IC50 calculated for the test fractions ranged from 0.134 to 0.808 mg/mL (Table 2). These values stay in agreement with the results reported elsewhere for amaranth protein hydrolysates [10,13]. On the other hand, values obtained in this work were lower than those found for other vegetable protein sources [14], or higher in comparison to buffalo milk [15]. Antioxidant potential based on the ABTS assay and expressed as SC50 varied from 0.992 to 6.931 mg/mL, and was slightly higher than that obtained for amaranth hydrolysates by in vitro digestion [16]. This may be explained by the fact that the type of bioactive peptides released depends not only on the protein nature, but also on the enzymes used in the particular case of hydrolysis [17,18]. Concerning the thrombin IC50 values, the results obtained were in the range of 0.992 to 38.46 mg/L, which stay in agreement with the values reported for amaranth hydrolyzed by alcalase and pepsin [12]. Finally, no statistically important differences were found for thrombin inhibition potential between H3 hydrolysate (alcalase and flavourzyme) investigated in this study, and amaranth hydrolyzed by alcalase and pepsin as reported in Sabbione et al. [12], which was also obtained via sequential two-enzyme process.

Based on the obtained results, fractions 22 and 45 were subjected to MALDI-TOF analysis (Figure 2), which allowed the identification of sequences with interesting multiple bioactivities.

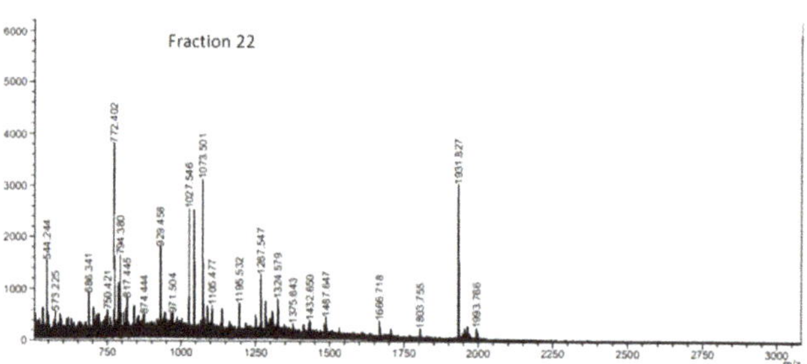

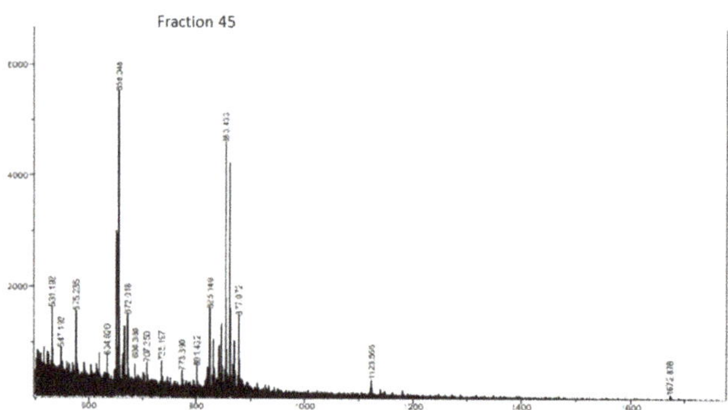

Figure 2. Fraction 22 and 45 MALDI-TOF mass Spectra.

The structure of the aminoacidic chains and the corresponding protein source are presented in Table 3.

Table 3. Peptides identified in the most active fractions from amaranth hydrolysate and their corresponding protein source.

Fraction	Mass *m/z*	Calc MH$^+$	Sequence	Protein
			Fraction 22	
34	1375.6435	1375.6341	ITASANEPDENKS	Agglutinin
3	573.2252	573.3516	LVRW	Agglutinin
16	874.4448	874.4813	NIDMLRL	Granule bound starch synthase I

Table 3. *Cont.*

Fraction	Mass *m/z*	Calc MH$^+$	Sequence	Protein
			Fraction 22	
12	794.3805	794.4203	RPVFEF	Granule bound starch synthase I
5	686.3414	686.4081	DPKLTL	Granule bound starch synthase I
3	573.2251	573.3617	IKEAL	Granule bound starch synthase I
13	812.3607	812.4265	NVEVHKS	Cystatin
			Fraction 45	
27	853.4330	853.4329	HVQLGHY	Agglutinin
14	707.3505	707.3212	SQIDTGS	Agglutinin
14	707.3502	707.3185	NWACTL	Agglutinin
4	547.1921	547.2997	VRWS	Agglutinin
29	861.3847	861.4299	CIHNIVY	Granule bound starch synthase I
26	845.4098	845.4254	EGTESIPL	Granule bound starch synthase I
24	841.4242	841.3841	PRYDQY	Granule bound starch synthase I
19	823.4281	823.3696	MSNIDML	Granule bound starch synthase I
13	686.3805	686.4080	DPKLTL	Granule bound starch synthase I
6	619.2805	619.3566	IPSRF	Granule bound starch synthase I
3	531.1927	531.3042	ARVW	Granule bound starch synthase I
2	505.1907	505.2447	CQAAL	Granule bound starch synthase I
1	503.1730	503.2715	EELL	Granule bound starch synthase I
1	503.1731	503.2823	LGVAGS	Granule bound starch synthase I

3. Discussion

Results obtained for the individual enzymatic hydrolysis were similar to those reported by Zhuang et al. [19] and Ma et al. [20], where it was observed that the enzymatic activity of alcalase and flavourzyme decreased after 90–120 min of the reaction course. The duration of enzymatic activity depends on the nature of the protein matrix and the amount of enzyme and substrate present in the medium. Additionally, other studies indicated a higher hydrolytic degree when only alcalase was used [21]. For H3 results, they are similar to those reported by Cumby et al. [22], who found that combined enzymatic hydrolysis increased peptide concentration when compared to hydrolysis performed solely with alcalase. It was also stated that hydrolysis with flavourzyme did not contribute to reach higher concentrations of free amine groups when alcalase was used in the first step of the combined enzymatic hydrolysis. On the other hand, the degree of hydrolysis depends on nature of protein and the specificity of the enzymes used [17,18].

Alcalase and flavourzyme have been used individually and in continuous hydrolytic processes applied to fragmentation of different food proteins in search of platelet inhibitory peptides [23,24]. In the case of amaranth proteins, the addition of a second enzyme different to flavourzyme, following hydrolysis with alcalase, has proven to enhance the inhibition potential of thrombin. This finding implies that the higher degree of hydrolysis, the greater bioactivity achieved [12,25]. In addition, Sabbione et al. [12] observed that the degree of hydrolysis of amaranth proteins, such as albumin and globulin, was an important factor in improved thrombin inhibition, reaching approximately 81%, which matches with the results obtained in present study (80% for H2, and 90% for H1 and H3).

The increase in thrombin inhibition activity may be due to the fact that amaranth protein has peptide sequences capable of inhibiting fibrinogen. These peptides are released by the action of proteolytic enzymes. Furthermore, it was observed that not only was the size of the peptide important for thrombin inhibition, but also the amino acid sequence, which might be homologous to the fibrinopeptides A and B from human fibrinogen (GGGVR-GP and PPSAR-GH, respectively) [26]. Therefore, antithrombotic activity is affected by the competition for platelet receptors between casoplatelin and the γ-chain of human fibrinogen (HHLGGAKQAGDV) [27,28], which is implicated in one of the three steps in thrombosis cascade reactions. Additionally, alcalase has been reported as an enzyme capable of releasing antithrombotic peptides form proteins like peanut [18], but no reports have been available so far on hydrolytic potential of flavourzyme to release bioactive peptides from amaranth.

In the case of antihypertensive activity, the results obtained in this study are similar to those reported by Ambigaipalan et al. [29]. They observed that the combination of two or more enzymes increased the inhibitory activity of angiotensin, converting the enzyme in date seed protein in a sequential enzymatic process. Many authors have reported that the variation in antihypertensive activity of the released peptides could be attributed to the differences in composition and hydrophobicity of the protein primary structure [30,31]. In this sense, hydrophobic residues of amino acids (leucine, valine, alanine, tryptophan, tyrosine, proline, and phenylalanine) bind at the ACE catalytic sites, acting as competitive inhibitors [32]. Nevertheless, milk proteins have been reported as a better source of bioactive peptides, especially those with antihypertensive activity reaching over 80% of ACE inhibition [33,34]. Furthermore, studies have been focused on peptides released from vegetable proteins, such as amaranth. In particular, 11S globulin was reported to show IC50 value in the ACE assay, ranging from 6.32 mM to 175 µM [9,35]. It has been known that the low molecular weight of peptides is a prerequisite for their antihypertensive activity [36]. In this study, peptides with the latter feature were identified using SDS-PAGE (data not shown), revealing typical weights of these biopeptides.

In addition to antihypertensive activity, peptide size (peptide chain length less than 20 amino acids) has been related to antioxidant capacity, showing that smaller peptides have greater potential [37]. This observation could explain the higher antioxidant activity obtained for H3 hydrolysate, which contained a higher concentration of free amine groups. In conclusion, during the course of enzymatic hydrolysis H1, H2, and H3 carried out in this study, peptides with different biological properties were released—for example, those able to chelate reactive agents, or donate electrons or hydrogen [38]. Moreover, in several studies in which sequential hydrolysis of proteins with different enzymes was performed, it has been observed that the antioxidant activity remained the same or might be increased by the addition of a second enzyme [39,40]. Thus, the application of two hydrolytic enzymes allows the release of new bioactive peptides, which could affect the antioxidant capacity of the hydrolysates. Consequently, it has been observed that protein hydrolysates might present higher antioxidant properties as they increase their content of small peptides [41].

For RP-HPLC peptide separation, the results obtained in this work are similar to those reported by Moronta et al. [42], who hydrolyzed amaranth proteins with alcalase. It was found that peptides with a higher content of polar amino acids showed higher anti-inflammatory activity. Nonetheless, a fraction found in the non-polar region of the chromatographic analysis (45 min) exhibited higher potential in terms of antihypertensive, antithrombotic, and antioxidant activity than other polar fractions. In some studies, it has been demonstrated that peptides containing hydrophobic amino acids might enhance biological capacities, such as antihypertensive and antioxidant activities [43].

As observed in Table, fraction 22 contained the longest peptide sequence, identified as ITASANEPDENKS, with a molecular weight of approximately 1.44 kDa, being the highest molecular weight found in both fractions. In the case of the peptide NIDMLRL, the last part of its sequence (-LRL) has been identified in silico as an inhibitor of ACE in amaranth. This asseveration may be explained by the presence of leucine as hydrophobic amino acid interacting with the active site of ACE [44].

Additionally, the LVRW sequence was found in fraction 22. The three amino acid chain LVR was previously described as a bioactive peptide with antihypertensive activity in fig sap having an IC50 lower than 20 µM [45]. Additionally, W might be also a bonding amino acid in the active site of ECA. On the other hand, the IC50 of ACE inhibitory potential found for fraction 45 was lower than for fraction 22. Fraction 45 contained VRWS, and the dipeptide VR was described as an antihypertensive agent by itself, with an IC50 of 52.80 µM [46]. In addition, even though it was bound to other amino acids, such as Y or SP, it retained its bioactivity [47,48]. Similarly, the presence of Tyr in the C-terminal end should promote binding to ACE and thus enhance its inhibition [49]. It was proved for the tripeptide

IVY [50], also present in the sequence CIHNIVY in fraction 45. Finally, in this fraction, various peptides containing Tyr residue were found.

In previous works on antioxidant peptides, it was stated that peptides with a length between 5 and 16 amino acids showed antiradical activity [51]. The results obtained in the present study support these findings, since the potential of antioxidant peptides had a molecular weight in the range of 500–1400 Da and contained 4–13 amino acid residues (Table 3).

One of the peptides described in fraction 22 was LVRW, which could play a role as an antioxidant agent. Haung et al. [52] determined that the presence of the RW amino acid sequence in the C-terminal end of the polypeptide chain might be accounted for by its high antiradical activity. A similar effect was observed for the peptide ARVW, were the antioxidant activity of Trp is mainly due to its indole group [53].

In fraction 45, the DPKLTL sequence was identified, which might have antioxidant capacity owning to the presence of the DPK fragment (previously described as an antioxidant peptide). Its bioactivity could be explained by the presence of aspartic acid, which has the ability to donate electrons and hydrogen. On the other hand, this fragment is bound to hydrophobic amino acids, namely Pro and Leu, which in turn could enhance the radical scavenging abilities of this peptide [54]. Additionally, the DPKLTL peptide might exhibit antithrombotic activity due to the presence of three amino DPK acid residues, previously known for inactivating thrombin in its active site [25].

According to the work of Wang et al. [55], peptides with antithrombotic bioactivity usually contain 3–20 amino acid residues. Table 3 shows that the longest peptide identified in fraction 22 had a 13-amino acid chain. Moreover, it has been reported that sequences containing Val and Tyr might possess antithrombotic activity [23]. These amino acids were mainly found among peptide sequences present in fraction 45. On the other hand, Pro-Arg bonding N-terminal end of polypeptide chains has been reported as a thrombin inhibitor in its active site [26]. This bond is present in the structure of PRYDQY; however, more exhaustive studies are necessary on antithrombotic peptides as thrombin contain three main structural domains (a catalytic site and two exosites (I and II)), and enzyme inhibition can take place to different extents at any of these sites [26,56].

4. Materials and Methods

4.1. Sample and Treatments

Raw *Amaranth hypochondriacus* spp. seeds were obtained from Xochimilco, Mexico City, in July 2016. Seeds were ground in a Chopin mill, segregated according to their molecules size, and only a fraction with molecule size between 200 and 800 μm was selected for the experiments. Protein extraction was performed following the methodology described by Martínez and Añón [57] with some modifications. First, flour was defatted with n-hexane (10% *w/v*) for 24 h. Then it was suspended in deionized water 10% (*w/v*), and pH 9 was adjusted with NaOH. The crude was incubated for 30 min at room temperature and centrifuged during 20 min at 10,000 rpm. The soluble protein fraction found in the supernatant was precipitated by pH 5 adjustment, using HCl. The sample was centrifuged 20 min at 10,000 rpm, and the pellet pH was adjusted to 7. Following the lyophilization, the protein extract was stored at 4 °C until further use. This protein was called protein extract.

4.2. Enzymatic Hydrolysis

The enzymatic hydrolysis was performed according to the method described by Tironi and Añón [58]. Briefly, 5 g of the protein extract was diluted in 100 mL of deionized water (milli Q 18.2 MΩ*cm, Manufacturer, Darmstadt, Germany). For alcalase hydrolysis (H1), prior to the reaction, the pH of the solution was adjusted to 10, and alcalase (≥2.4 U/g, Anson Units; Sigma-Aldrich (St. Loui, MO, USA) was added at concentration of 8 μL/100 mg of the sample. For flavourzyme hydrolysis (H2), the pH of the reaction media was adjusted to 7 and flavourzyme (≥500 U/g; Sigma Aldrich, St. Louis, MO, USA) was added at concentration of 5 μL/100 mg of the sample. Finally, when both enzymes were used in a two-step continuous hydrolytic process (H3), after 2 h of hydrolysis with alcalase, the enzyme was inactivated by heating to 85 °C for 10 min. The solution was adjusted to pH 7 and hydrolysis with flavourzyme was carried out. For all three methods, hydrolysis was followed up during 4 h and progress of the reaction was monitored every 20 min. The aliquots were heated to 85 °C during 10 min and frozen until further use.

4.3. Free Amine Groups Analysis by TNBS Test

In order to determine free amine groups released during the enzymatic hydrolysis of the amaranth proteins, the 2,4,6-Trinitrobenzenesulfonic acid (TNBS) test was performed, with some modifications, according to the protocol described by Sashidhar et al. [59]. Briefly, 5% TNBS solution (Sigma-Aldrich, (St. Loui, MO, USA) was diluted in 0.21 M phosphate buffer (pH 8.2) to a final concentration of 1% (*v/v*). Two milliliters of the prepared substrate was added to 2 mL of phosphate buffer (0.21 M; pH 8.2), and 0.25 mL of the test sample. The mixture was incubated during 1 h at 50 °C and the reaction was stopped by the addition of 2 mL of 0.1 N HCl. The absorbance was read at 340 nm. Results were plotted against a calibration curve prepared by different glycine concentrations (0, 0.05. 0.1, 0.15, 0.2, and 0.25 mg/mL) using the following equation (R = 0.9972):

$$y = 0.004x + 0.134 \tag{1}$$

Results are expressed as milligrams of free amines per liter (mg/L NH:$^-$).

4.4. Antihypertensive Activity

Inhibitory effect on angiotensin converting enzyme (ACE) was evaluated spectrophotometrically according to the method of Cushman et al. [60] using Hippuril-Histidyl-Leucine (St. Loui, MO, USA) as substrate. Briefly, 5 mM substrate solution was prepared in 0.1 M sodium borate buffer pH 8.3 containing 0.3 M sodium chloride. A 5 mM 100 µL aliquot of substrate solution was mixed with 40 µL of the test sample before adding 10 µL of angiotensin converting enzyme (EC 3.4.15.1, 5.1 U/mg; Sigma-Aldrich). The reaction mixture was incubated during 1 h 15 min at 37 °C and the enzyme was inactivated with 1 mL of 0.1 M HCl. The hippuric acid formed in this reaction was extracted with ethyl acetate, concentrated under reduced pressure, and finally re-dissolved in distilled water. The absorbance was measured at 220 nm in a GENESYS spectrometer.

The inhibitory activity of ACE was calculated using the formula:

$$\% \text{ inhibitory activity} = (AbsC - AbsM)/(AbsC - AbsB) \times 100 \tag{2}$$

where:
    AbsC: Hippuric acid formed during the reaction with ACE without inhibitor.
    AbsB: Hippuril-Histidyl-Leucine that did not react and was extracted with ethyl acetate.
    AbsM: Hippuric acid formed after the reaction with ACE in the presence of inhibitory substance.

4.5. Antithrombotic Activity

To evaluate antithrombotic activity, the methodology developed by Zhang et al. [61] was followed, including modifications of sodium chloride concentration proposed by Pérez-Escalante et al. [56]. Absorbance at 405 nm was measured before adding the enzyme to a microplate reader (Power Wave XS UV-Biotek, software KC Junior, USA) and after 10 min of incubation with the enzyme at 37 °C. The percentage of inhibition (% inhibition) was calculated following the equation:

$$\% \text{ inhibitory activity} = [(C - CB) - (S - SB)]/(C - CB) * 100 \tag{3}$$

where:
    CB (control blank): The initial absorbance of the negative control of inhibition.

C (control): The absorbance of the negative control after 10 min of incubation with thrombin.
SB (sample blank): The initial absorbance of the sample.
S (sample): The absorbance of the sample after 10 min of incubation with thrombin.

4.6. Antioxidant Activity

Antioxidant activity was evaluated by three different methods.

4.6.1. ABTS test

Radical scavenging capacity was measured using the radical cation 2,2′-azino-bis 3-ethylbenzothiazoline-6-sulphonic acid (ABTS•+), which was produced by mixing 7 mM of ABTS•+ stock solution with 2.45 mM potassium persulfate in the dark at room temperature for 16 h prior to its usage. The ABTS•+ solution was diluted with deionized water, so that absorbance measured at 754 nm was of 0.70 ± 0.02. An aliquot of 20 μL of test sample was added to 980 μL of the diluted ABTS•+ solution, and after 7 min of incubation at room temperature, the absorbance readings were taken at 754 nm in a microplate reader (Power Wave XS UV-Biotek, soft-ware KC Junior, USA). Antioxidant capacity was expressed as milligrams Trolox equivalents per liter (mg TE/100 g) [62].

4.6.2. DPPH Test

Radical scavenging activity was measured using 2,2-diphenyl-1-picrylhydrazyl (DPPH•) radical. An ethanolic solution (7.4 mg/100 mL) of the stable DPPH• radical was prepared. Then, 100 μL of the test sample was added to 500 μL of DPPH• solution, and it was left to sit at room temperature for 1 h. The solution was stirred and centrifuged at 3000 rpm for 10 min. Finally, absorbance of the supernatant was measured at 520 nm in a microplate reader. Antioxidant activity was expressed as micromole of Trolox equivalents per liter (TE μmol/100 g) [63].

4.6.3. FRAP Test

FRAP antioxidant activity was evaluated according to the Benzie and Strein method [64], in which 0.3 M sodium acetate buffer pH 3.6, TPTZ, 20 mM FeCl3, and 5M FeSO4 were elaborated. For the preparation of FRAP buffer, TPTZ and $FeCl_3$ were mixed at 10:1 ratio (*v:v*). Briefly, 30 μL of test sample was mixed with 900 μL of FRAP solution and 90 μL of distilled water, and agitated. The mixture was incubated during 10 min at 37 °C, after which the absorbance was measured at 593 nm in a microplate reader. Results were compared with a calibration curve constructed for $FeSO_4$ standards at concentrations ranging from 0 to 1000 mM. Antioxidant activity was expressed as micromole equivalents of Fe (II) per 100 g (μmolEFeII/100 g).

4.7. Identification of Bioactive Peptides by RP-HPLC

4.7.1. Sample Preparation

For sample preparation, 10 mg/mL of freeze-dried amaranth protein hydrolysates were prepared with phosphate buffer (pH 7.8). The crude mixture was stirred during 1 h at 37 °C and followed by centrifugation at 10,000 rpm for 10 min at room temperature. The resulting supernatant with corresponding soluble fractions was separated and stored.

4.7.2. RP-HPLC Separation

Peptides were separated by reversed-phase chromatography on a HPLC (Waters, USA) system equipped with a C8 column (250 mm × 4.6 mm × 5 mm; Waters) and photodiode array detector (Spectra System Thermo Scientific, USA) A gradient elution was applied from 100 to 0% A in 56 min at flow rate of 1 mL/min. A binary mobile phase consisted of solvent A: 0.065% trifluoroacetic acid [TFA] in water/acetonitrile [ACN] 98:2, and solvent B: 0.065% TFA in water/ACN 35:65. Detection was performed at 280 nm and temperature was set at 40 °C. The injection volume was 200 μL. Fractions were manually collected every 1 min in Eppendorf tubes and protein content was evaluated by the Bradford assay. Finally, samples were lyophilized (Labconco DrySystem/freezone 4.5) and bioactivity analysis was carried out only on the lyophilized samples containing protein.

4.7.3. MALDI-TOF Spectrometry

The collected fractions with the highest bioactivities were filtered through a Minisart RC4 filter (0.45 μm) and analyzed by a matrix-assisted laser desorption ionization (MALDI) mass spectrometer, equipped with a delayed extraction source and a 355 nm pulsed nitrogen laser. A MALDI scoutMTPTM was run in the linear mode. A 100-times diluted sample was mixed with 1 volume of 20 mg/mL of sinapinic acid in acetonitrile/water 50:50 (*v/v*). Finally, 0.5 μL of the mixture was deposited onto the MALDI target plate. All spectra were the result of signal averaging of 200 shots. The MALDI-TOF/TOF MS/MS was run in the positive refractor mode. The peptide sequencing was performed by processing the MS/MS spectra using Auto eXecute software. Mascot (Matrix Science Inc., Boston, MA, US) software was used to identify and characterize peptide structures.

4.8. Statistical Analysis

All data were obtained in triplicate and expressed as mean ± standard deviation (SD). The data were analyzed by one-way analysis of variance (ANOVA) tests, and the differences among means were compared using the Tukey test with significance level set at $p < 0.05$, using the SPSS® System for WINTM version 15.0 (IBM®Armonk, New York, NY, USA).

5. Conclusions

Amaranth protein hydrolysate, obtained through enzymatic reaction with alcalase and flavourzyme in a sequential two-step hydrolytic process, may be a source of bioactive peptides. Sequences were different from those obtained in more commonly performed enzymatic hydrolysis using pepsin and pancreatin for amaranth proteins digestion. The biological activities identified for some peptide sequences in this study have been proven in other food sources. Moreover, novel amino acid chains with possible multi-functional activities were identified. This is the first report of the multiple bioactivities of peptide fractions derived from hydrolyzed amaranth proteins. Additionally, this study shows that amaranth hydrolyzed with alcalase and flavourzyme could be used in the nutraceutical industry, as a value-added ingredient with multi-functional bioactive properties.

Author Contributions: Conceptualization, L.G.G.-O. and A.A.-N.; methodology, L.G.G.-O., A.A.-N. and G.M.R.-S.; software, E.C.-L., P.R.-L. and A.C.-S.; validation, P.R.-L., A.C.-S. and L.G.G.-O.; formal analysis, L.G.G.-O., A.A.-N. and P.R.-L.; investigation, A.A.-N.; resources, A.C.-S. and L.G.G.-O.; data curation, E.C.-L. and A.C.-S.; writing—original draft preparation, A.A.-N. and L.G.G.-O.; writing—review and editing, A.A.-N. and L.G.G.-O.; visualization, L.G.G.-O.; supervision, L.G.G.-O. and G.M.R.-S.; project administration, L.G.G.-O.; funding acquisition, L.G.G.-O., E.C.-L. and G.M.R.-S.

Funding: This research received no external funding.

Conflicts of Interest: The authors declare no conflict of interest.

Appendix A

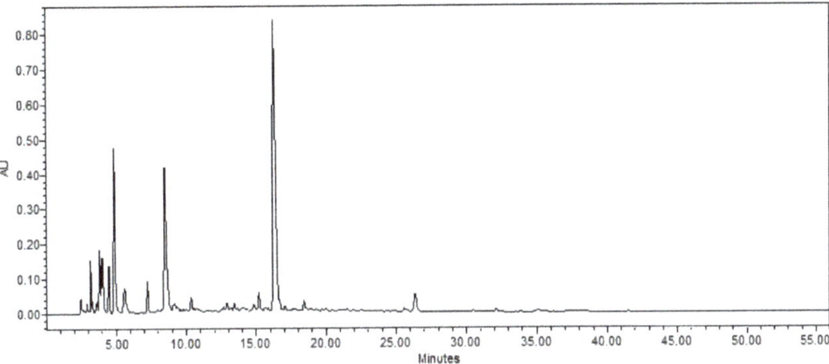

Figure A1. RP-HPLC separation of amaranth hidrolizated fractions.

References

1. Jew, S.; AbuMweis, S.S.; Jones, P.J.H. Evolution of the human diet: Linking our ancestral diet to modern functional foods as a means of chronic disease prevention. *J. Med. Food* **2009**, *12*, 925–934. [CrossRef] [PubMed]
2. Udenigwe, C.C.; Aluko, R.E. Food protein-derived bioactive peptides: Production, processing, and potential health benefits. *J. Food Sci.* **2012**, *77*, 11–24. [CrossRef] [PubMed]
3. Mazorra-Manzano, M.A.; Ramírez-Suarez, J.C.; Yada, R.Y. Plant proteases for bioactive peptides release: A review. *Crit Rev. Food Sci. Nutr.* **2018**, *58*, 2147–2163. [CrossRef] [PubMed]
4. Wattanasirithama, L.; Theerakulkaita, C.; Wickramasekara, S.; Maierb, C.S.; Stevensc, J.F. Isolation and identification of antioxidant peptides from enzymatically hydrolyzed rice bran protein. *Food Chem.* **2016**, *192*, 156–162. [CrossRef] [PubMed]
5. Najafian, L.; Babji, A.S. Production of bioactive peptides using enzymatic hydrolysis and identification antioxidative peptides from patin (Pangasius sutchi) sarcoplasmic protein hydrolysate. *J. Funct. Foods* **2014**, *9*, 280–289. [CrossRef]
6. Baltia, R.; Bougatef, A.; Silaa, A.; Guillochon, D.; Dhulster, P.; Nedjar-Arroume, N. Nine novel angiotensin I-converting enzyme (ACE) inhibitory peptides from cuttlefish (Sepia officinalis) muscle protein hydrolysates and antihypertensive effect of the potent active peptide in spontaneously hypertensive rats. *Food Chem.* **2015**, *170*, 519–525. [CrossRef] [PubMed]
7. Raikos, V.; Dassios, T. Health-promoting properties of bioactive peptides derived from milk proteins in infant food: A review. *Dairy Sci. Technol.* **2014**, *94*, 91–101. [CrossRef] [PubMed]
8. Tovar, L.R. Biotechnology for an Ancient Crop- Amaranth. In *Amaranth Biology, Chemistry and Technology*, 1st ed.; Paredes-López, O., Ed.; CRC Press: Boca Raton, FL, USA, 2018; Volume 1, pp. 14–36.
9. Mendonça, S.; Saldiva, P.H.; Cruz, R.J.; Areas, A.G. Amaranth protein presents cholesterol-lowering. *Food Chem.* **2009**, *116*, 738–742. [CrossRef]
10. Tovar-Pérez, E.G.; Guerrero-Legarreta, B.C.; Ferrés-González, A.; Soriano-Santos, J. Angiotensin I-converting enzyme-inhibitory peptide fractions from albumin 1 and globulin as obtained of amaranth grain. *Food Chem.* **2009**, *116*, 437–444. [CrossRef]
11. Orsini-Delgado, M.C.; Torini, V.A.; Añón, M.C. Antioxidant activity of amaranth protein or their hydrolysates under simulated gastrointestinal digestion. *Food Sci. Technol.* **2011**, *44*, 1752–1760. [CrossRef]
12. Sabbione, A.C.; Scilingo, A.; Añón, M.A. Potential antithrombotic activity detected in amaranth proteins and its hydrolysates. *Food Sci. Technol.* **2015**, *60*, 171–177. [CrossRef]
13. Tiengo, A.; Faria, M.; Neitto, E.M. Characterization and ACE-Inhibitory activity of Amaranth proteins. *J. Food Sci.* **2009**, *74*, H121–H126. [CrossRef] [PubMed]
14. Rudolph, S.; Lunow, D.; Kaiser, S.; Henle, T. Identification and quantification of ACE-inhibiting peptides in enzymatic hydrolysates of plant proteins. *Food Chem.* **2017**, *224*, 19–25. [CrossRef] [PubMed]

15. Abdel-Hamid, M.; Otte, J.; De Gobba, C.; Osman, A.; Hamad, E. Angiotensin I-converting enzyme inhibitory activity and antioxidant capacity of bioactive peptides derived from enzymatic hydrolysis of buffalo milk proteins. *Int. Dairy J.* **2017**, *66*, 91–98. [CrossRef]
16. Orsini-Delgado, A.; Galleano, M.; Añón, M.C.; Tironi, V.A. Amaranth peptides from simulated gastrointestinal digestión: Antioxidant Activity against Reactive species. *Plants Foods Human Nutr.* **2015**, *70*, 27–34. [CrossRef]
17. Zhang, J.; Zhang, H.; Wang, L.; Guo, X.; Wang, G.; Wang, X.; Yao, H. Isolation and identification of antioxidative peptides from rice endosperm protein enzymatic hydrolysates by consecutive chromatography and MALDI-TOF/TOF MS/MS. *Food Chem.* **2010**, *119*, 226–234. [CrossRef]
18. Zhang, B.S. In vitro antithrombotic activities of peanut protein hydrolysates. *Food Chem.* **2016**, *202*, 1–8. [CrossRef]
19. Zhuang, H.; Tang, N.; Yuan, Y. Purification and identification of antioxidant peptides from corn gluten meal. *J. Funct. Foods* **2013**, *5*, 1810–1821. [CrossRef]
20. Ma, Y.; Wang, L.; Sun, X.; Zhang, J.; Wang, J.; Li, Y. Study on Hydrolysis Conditions of Flavourzyme in Soybean Polypeptide Alcalase Hydrolysate and Soybean Polypeptide Refining Process. *Adv. J. Food Sci. Technol.* **2014**, *6*, 1027–1032. [CrossRef]
21. Foh, M.B.; Amadou, I.; Foh, B.M.; Kamara, M.T.; Xia, W. Functionality and antioxidant properties of Tilipa (Oreochromis niloticus) as influences by degree of hydrolysis. *Int. J. Mol. Sci.* **2010**, *11*, 1851–1869. [CrossRef]
22. Cumby, N.; Zhong, Y.; Naczk, M.; Shahidi, F. Antioxidant activity and water-holding capacity of canola protein hydrolysates. *Food Chem.* **2008**, *109*, 144–148. [CrossRef]
23. Cian, R.E.; Garzón, A.G.; Martínez-Augustin, O.; Botto, C.C.; Drago, S.R. Antithrombotic activity of Brewer's spent grain peptides and their effects on blood coagulation pathways. *Plant. Foods Hum. Nutr.* **2018**, *73*, 241–246. [CrossRef]
24. Kasiwut, J.; Sirinupong, N.; Wirote, Y. The anticoagulant and angiotensin I-Converting Enzyme (ACE) Inhibitory peptides from tuna cooking juice produced by alcalase. *Curr. Nutr. Food Sci.* **2018**, *14*, 225–234. [CrossRef]
25. Sabbione, A.C.; Nardo, E.A.; Añón, M.C.; Scilingo, A. Amaranth peptides with antithrombotic activity released by sumulated gastrointestinal digestion. *J. Funct. Foods* **2016**, *20*, 204–214. [CrossRef]
26. Cheng, S.; Tu, M.; Liu, H.; Zhao, G.; Du, M. Food-derived antithrombotic peptides: Preparation, identification and interactions with thrombin. *Crit. Rev. Food Sci. Nutr.* **2019**, *2019*. [CrossRef] [PubMed]
27. Laudano, A.P.; Doolittle, R.F. Synthetic peptides derivatives that bind to fibrinogen and prevent the polymerization of fibrin monomers. *PNAS USA* **1978**, *75*, 3085–3089. [CrossRef] [PubMed]
28. Fiat, A.M.; Levy-Toledano, S.; Caen, J.P.; Jolles, P. Biologically active peptides of casein and lactoferrin implicated in platelet function. *J. Dairy Res.* **1989**, *56*, 351–355. [CrossRef] [PubMed]
29. Ambigaipalan, P.; Al-Khalifa, A.S.; Shahidi, F. Antioxidant and angiotensin I converting enzyme (ACE) inhibitory activities of date seed protein hydrolysate prepared using Alcalase, Flavourzyme and Thermolysin. *J. Funct. Foods* **2015**, *18*, 1125–1137. [CrossRef]
30. Jung, W.K.; Mendis, E.; Je, J.; Park, P.J.; Son, B.; Kim, H.C.; Choi, Y.K.; Kim, S.K. Angiotensin I-converting enzyme inhibitory peptide from yellowfin sole (Limanda aspera) frame protein and its antihypertensive effect in spontaneously hypertensive rats. *Food Chem.* **2006**, *94*, 26–32. [CrossRef]
31. Wijesekara, I.; Kim, S.K. Angiotensin-I-converting enzyme (ACE) inhibitors from marine resources: Prospects in the pharmaceutical industry. *Mar. Drugs* **2010**, *8*, 1080–1093. [CrossRef] [PubMed]
32. Hanafi, M.A.; Hashim, S.N.; Yea, C.S.; Ebrahimpour, A.; Zarei, M.; Muhammad, K.; Abdul-Hamid, A.; Saari, N. High angiotensin-I converting enzyme (ACE) inhibitory activity of Alcalase-digested green soybean (Glycine max) hydrolysates. *Food Res. Int.* **2018**, *106*, 589–597. [CrossRef] [PubMed]
33. González-González, C.R.; Tuohy, K.M.; Jauregi, P. Production of angiotensin-I-converting enzyme (ACE) inhibitory activity in milk fermented with probiotic strains: Effects of calcium, pH and peptides on the ACE-inhibitory activity. *Int. Dairy J.* **2011**, *21*, 615–622. [CrossRef]
34. Ayyash, M.; Al-Dhaheri, A.S.; Mahadin, S.A.; Kizhakkayil, J.; Abushelaibi, A. In vitro investigation of anticancer, antihypertensive, antidiabetic and antioxidant activities of camel milk fermented with camel milk probiotic: A comparative study with fermented bovine milk. *J. Dairy Sci.* **2018**, *101*, 1–12. [CrossRef] [PubMed]
35. Vecchi, B.; Añón, M.C. ACE inhibitory tetrapeptides from Amaranthus hypochondriacus 11S globulin. *Phytochemistry* **2009**, *70*, 864–870. [CrossRef] [PubMed]

36. Miralles, B.; Amigo, L.; Recio, I. Critical Review and Perspective on Food-Derived Antihypertensive Peptides. *J. Agric. Food Chem.* **2018**, *66*, 9384–9390. [CrossRef] [PubMed]

37. Zou, T.B.; He, T.P.; Li, H.B.; Tang, H.W.; Xia, W.Q. The structure-activity relationship of the antioxidant peptides from natural proteins. *Molecules* **2016**, *12*, 72. [CrossRef]

38. Virtanen, T.; Pihlanto, A.; Akkanen, S.; Korhonen, H. Development of antioxidant activity in milk whey during fermentation with lactic acid bacteria. *J. Appl. Microbiol.* **2007**, *102*, 106–115. [CrossRef] [PubMed]

39. Je, J.Y.; Lee, K.H.; Lee, M.H.; Ahn, C.B. Antioxidant and antihypertensive protein hydrolysates produces from tuna liver by enzymatic hydrolysis. *Food Res. Int.* **2009**, *42*, 1266–1272. [CrossRef]

40. Chirinos, R.; Ochoa, K.; Aguilar-Galvez, A.; Carpentier, S.; Pedreschi, R.; Campos, D. Obtaining of peptides with in vitro antioxidant and angiotensin I converting enzyme inhibitory activities from cañihua protein (Chenopodium pallidicaule Aellen). *J. Cereal. Sci.* **2018**, *83*, 139–146. [CrossRef]

41. Choi, Y.; Lim, T.; He, Y.; Hwang, T. Chemical characteristics and antioxidant properties of wheat gluten hydrolysates produced by single and sequential enzymatic hydrolyses using commercial proteases and their application in beverage systems. *J. Food Meas. Charact.* **2019**, *13*, 745–754. [CrossRef]

42. Moronta, J.; Smaldini, P.L.; Docena, G.H.; Añón, M.C. Peptides of amaranth were targeted as containing sequences with potential anti-inflammatory properties. *J. Funct. Foods* **2018**, *21*, 463–473. [CrossRef]

43. Ijaritomi, O.S.; Malomo, S.A.; Alashi, A.M.; Nwachukwu, I.D.; Fagbemi, T.N.; Osundahunsi, O.F.; Aluko, R.E. Antioxidant and antihypertensive activities of wonderful cola (Buchholzia coriacea) seed protein and enzymatic protein hydrolysates. *J. Food Bioact.* **2018**, *3*, 133–143. [CrossRef]

44. Montoya-Rodríguez, A.; Gómez-Fávela, M.A.; Reyes-Moreno, C.; Milán-Carrillo, J.; González de Mejía, E. Identification of Bioactive Peptides Sequences from Amaranth (Amaranthus hypochodriacus) Seeds Proteins and Their Potential Role in the Prevention of Chronic Diseases. *Compr. Rev. Food Sci. Food Saf.* **2015**, *14*, 139–159. [CrossRef]

45. Balgir, P.P.; Sharma, M. Biopharmaceutical potential of ACE-Inhibitory peptides. *J. Proteomics Bioinform.* **2017**, *10*, 171–177. [CrossRef]

46. Gómez-Ruiz, J.A.; Ramos, M.; Recio, I. Identification of a novel angiotensin-converting enzyme-inhibitory peptides ovine milk proteins by CE-MS and chromatographic techniques. *Electrophoresis* **2007**, *28*, 4202–4211. [CrossRef] [PubMed]

47. Wang, X.; Wu, S.; Xu, D.; Xie, D.; Guo, H. Inhibitor and substrate binding by angiotensin-converting enzyme: Quantum Mechanical/Molecular Mechanical Molecular Dynamics Studies. *J. Chem. Inf. Model.* **2011**, *51*, 1074–1082. [CrossRef] [PubMed]

48. De Gobba, C.; Tompa, G.; Otte, J. Bioctive peptides from caseins released by cold active proteolytic enzymes from Arsukibacterium ikkense. *Food Chem.* **2014**, *15*, 205–215. [CrossRef]

49. Wu, J.; Aluko, R.E.; Nakai, S. Structural Requirements of Angiotensin I-Converting Enzyme Inhibitory peptides: Quantitative Structure-Activity Reletionship Study of di and tripeptides. *J. Agric. Food Chem.* **2006**, *54*, 732–738. [CrossRef]

50. Ueno, T.; Tanaka, M.; Matsiu, T.; Matsumoto, K. Determination of antihypertensive small peptides, Val-Tyr and Ile-Val-Tyr, by fluorometric high-performance liquid chromatography combined with a double heart-cut column-switching technique. *Anal. Sci.* **2005**, *21*, 997–1000. [CrossRef]

51. Umayaparvathi, S.; Meenakshi, S.; Vimalraj, V.; Arumugam, M.; Sivagami, G.; Balasubramanian, T. Antioxidant activity and anticancer effect of bioactive peptide from enzymatic hydrolysate of oyster (Saccostrea cucullata). *Biomed. Prev. Nutr.* **2014**, *4*, 343–353. [CrossRef]

52. Haung, W.Y.; Majumder, K.; Wu, J. Oxygen radical absorbance capacity of peptuides from egg white protein ovotransferrin and their interaction with phytochemicals. *Food Chem.* **2010**, *123*, 635–641. [CrossRef]

53. Nimalaratne, C.; Lopes-Lutz, D.; Scheiber, A.; Wu, J. Free aromatic amino acids in egg yolk show antioxidant properties. *Food Chem.* **2011**, *129*, 151–161. [CrossRef]

54. Liu, C.; Ren, D.; Li, J.; Fang, L.; Wang, J.; Liu, J.; Min, W. Cytoprotective effect and purification of novel antioxidant peptides from hazelnut (C. heterophylla Fish) protein hydrolysates. *J. Funct. Foods* **2018**, *42*, 203–215. [CrossRef]

55. Wang, X.Q.; Yu, H.H.; Xing, R.; Li, P.C. Characterization, preparation, and purification of marine bioactive peptides. *Bio Med. Res. Int.* **2017**, *2017*, 1–16. [CrossRef] [PubMed]

56. Pérez-Escalante, E.; González-Olivares, L.G.; Cruz-Guerrero, A.E.; Galán-Vidal, C.A.; Páez-Hernández, M.E.; Álvarez-Romero, G.A. Size exclusion chromatography (SEC-HPLC) as an alternative to study thrombin inhibition. *J. Chromatogr. B* **2018**, *1074*, 34–38. [CrossRef] [PubMed]

57. Martínez, N.E.; Añón, C.M. Composition and structural characterization of amaranth protein isolates: An electrophoretic and calorimetric study. *J. Agric. Food Chem.* **1996**, *44*, 2523–2530. [CrossRef]

58. Tironi, V.A.; Añón, M.C. Amaranth proteins as a source of antioxidant peptides: Effect of proteolysis. *Food Res. Int.* **2010**, *43*, 315–322. [CrossRef]

59. Sashidhar, R.B.; Capoor, A.K.; Ramana, D. Quantification of amino group using amino acids as reference standards by trinitrobenzene sulfonic acid: A simple spectrophotometric method for the estimation of hapten to carrier protein ratio. *J. Inmunol. Methods* **1995**, *167*, 121–127. [CrossRef]

60. Cushman, D.W.; Cheung, H.B.; Sabo, E.F.; Ondetti, M.A. Design of potent competitive inhibitors of angiotensin converting enzyme. Carboxylalkanoyl and mercaptoalkanoyl aminoacids. *Biochemistry* **1977**, *16*, 5484–5491. [CrossRef] [PubMed]

61. Zhang, B.S.; Zhang, W.; Xu, S.Y. Antioxidant and antithrombotic activity of raoeseed peptides. *J. Am. Oil Chem. Soc.* **2008**, *85*, 521–527. [CrossRef]

62. Kuskoski, E.; Asuero, A.; Troncoso, A.; Mancini-Filho, J.; Fett, R. Aplicación de diversos métodos químicos para determinar actividad antioxidante en pulpa de fruto. *Ciênc Tecnol. Aliment.* **2005**, *25*, 726–732. [CrossRef]

63. Delgado-Andrade, C.; Rufian-Henares, J.A.; Morales, F.J. Assessing the antioxidant activity of melanoidins from coffee brews by different antioxidant methods. *J. Agric. Food Chem.* **2005**, *53*, 7832–7836. [CrossRef] [PubMed]

64. Benzie, I.; Strain, J. The Ferric Reducing Ability of Plasma (FRAP) as a measure if "Antioxidant power". The FRAP assay. *Anal. Biochem.* **1996**, *239*, 70–76. [CrossRef] [PubMed]

Sample Availability: Samples of the compounds are not available from the authors.

Article

Green Extraction of Phenolic Acids from *Artemisia argyi* Leaves by Tailor-Made Ternary Deep Eutectic Solvents

Li Duan [1,†], Chenmeng Zhang [1,†], Chenjing Zhang [1], Zijing Xue [2], Yuguang Zheng [2,*] and Long Guo [2,*]

1   College of Chemistry and Material Science, Hebei Normal University, Shijiazhuang 050024, China
2   School of Pharmacy, Hebei University of Chinese Medicine, Shijiazhuang 050200, China
*   Correspondence: zyg314@163.com (Y.Z.); guo_long11@163.com (L.G.); Tel.: +86-0311-89926316 (Y.Z.); +86-0311-89926017 (L.G.)
†   These authors contributed equally to this work.

Academic Editor: Jose M. Miranda
Received: 25 June 2019; Accepted: 3 August 2019; Published: 5 August 2019

Abstract: The *Artemisia argyi* leaf (AL) has been used as a traditional medicine and food supplement in China and other Asian countries for hundreds of years. Phytochemical studies disclosed that AL contains various bioactive constituents. Among bioactive constituents, phenolic acids have been recognized as the main active compounds in AL. To the best of our knowledge, no research has been focused on extraction method for the bioactive phenolic acids from AL. Nowadays, deep eutectic solvents (DESs) are emerging as a new type of green and sustainable solvent for efficient extraction of bioactive compounds from natural products. In the present study, an environmentally friendly extraction method based on DESs was established to extract bioactive phenolic acids from ALs. Diverse tailor-made solvents, including binary and ternary DESs, were explored for simultaneous extraction of four phenolic acids (3-caffeoylquinic acid, 3,4-di-*O*-caffeoylquinic acid, 3,5-di-*O*-caffeoylquinic acid, and 4,5-di-*O*-caffeoylquinic acid) from AL. The results indicated that the ternary DES composed of a 2:1:2 molar ratio of choline chloride, malic acid, and urea showed enhanced extraction yields for phenolic acids compared with conventional organic solvents and other DESs. Subsequently, the extraction parameters for the four phenolic acids by selected tailor-made DESs, including liquid–solid ratios, water content (%) in the DESs, and extraction time, were optimized using response surface methodology and the optimal extraction conditions were: extraction time, 23.5 min; liquid–solid ratio, 57.5 mL/g (mL of DES/g dry weight of plant material); water content, 54%. The research indicated that DESs were efficient and sustainable green extraction solvents for extraction of bioactive phenolic acids from natural products. Compared to the conventional organic solvents, the DESs have a great potential as possible alternatives to those organic solvents in health-related areas such as food and pharmaceuticals.

Keywords: deep eutectic solvents; *Artemisia argyi* leaves; phenolic acids; extraction; response surface methodology

---

1. Introduction

The *Artemisia argyi* leaf (AL), which is widely distributed in China and other Asian countries, has been used as a traditional medicine or food supplement for hundreds of years [1]. As a traditional Chinese medicine, AL is reported to possess antioxidant, antibacterial, anti-inflammatory, anticancer, hemostatic, and analgesic activities and is commonly used for treatment of hemorrhage, pain, eczema, and menstruation-related symptoms [2–4]. AL is also consumed as a food ingredient because of its delicious flavor and distinctive smell. In China, AL is used as a common condiment and colorant for the traditional Chinese food "Qingtuan". In Japan, AL is added into food as an additive to enhance the flavor and nutrition [5].

Although the AL has been used as an herbal medicine and food ingredient for a long time, studies on its bioactive compositions are still limited. Phytochemical studies disclosed that AL contains various bioactive constituents, mainly including volatile oils, phenolic acids, flavonoids, and terpenoids [6,7]. Among bioactive constituents, phenolic acids have been recognized as the main active compounds in AL [8,9]. Therefore, the content of the phenolic acids is an important index for quality analysis and normal applications of AL. To date, several bioactive phenolic acids, such as 3-caffeoylquinic acid, 3,4-di-O-caffeoylquinic acid, 3,5-di-O-caffeoylquinic acid, and 4,5-di-O-caffeoylquinic acid, have been isolated from AL. A number of analytical methods, including high-performance liquid chromatography (HPLC) and high-performance liquid chromatography coupled with mass spectrometry (HPLC-MS), have been used for qualitative and quantitative analysis of the main phenolic acids in AL [5,7]. However, the research focused on extraction methods for bioactive phenolic acids from AL is still limited.

Nowadays, conventional organic solvents, such as alcohols, ethyl acetate, acetone, and chloroform, are widely used in the extraction of bioactive components from natural sources [10]. However, the consumption of large amounts of these volatile and hazardous organic solvents may contribute to environmental pollution and leave unacceptable solvent residues in extracts. Therefore, in analytical chemistry, green extraction methods which are environmentally friendly and sustainable for sample preparation have received more and more attention [11]. Since being introduced as a new type of green solvent, deep eutectic solvents (DESs) have rapidly gained great interest as sustainable alternatives to conventional organic solvents. DESs are prepared by simply mixing two or more naturally occurring, inexpensive, and biodegradable components together to obtain a eutectic mixture [12]. The availability, low cost, biodegradability, and environmental friendliness of the components make the DESs versatile alternatives to conventional organic solvents [13]. Due to their excellent properties, including biodegradability, low toxicity, solute stabilization, and low cost, DESs have been widely used in organic synthesis, separation processes, and biomedical applications [14,15].

Generally, DESs are prepared by simply mixing two or more naturally occurring, inexpensive, and biodegradable components together to obtain eutectic mixtures. As tunable solvents, diverse possible combinations of starting components have different targeted functionality, which means that we can increase the solubility and extraction efficiency of DESs for target compounds by selecting appropriate combinations of starting components. The tailor-made DESs have tremendous potential for efficient and simultaneous extraction and separation of compounds which have obvious differences in nature [16]. Recently, many reports have shown that tailor-made DESs were successfully employed in the extraction and separation of different kinds of bioactive compounds, such as phenolic acids, flavonoids, alkaloids, and saponins, from various plant materials [17–22]. Nonetheless, most of the research used tailor-made binary DESs as extraction solvents for bioactive compound extractions, the number of reports on the application of tailor-made ternary DESs for extraction is still limited, and the efficiency of DESs for extraction of bioactive phenolic acids from AL still remains unknown.

In the present study, in order to evaluate DESs for the extraction of phenolic acids, several tailor-made binary and ternary DESs were used for simultaneous extraction of four bioactive phenolic acids (3-caffeoylquinic acid, 3,4-di-O-caffeoylquinic acid, 3,5-di-O-caffeoylquinic acid, and 4,5-di-O-caffeoylquinic acid) from AL, and the extraction efficiency of tailor-made DESs was compared with that of conventional organic solvents. Moreover, the extraction parameters for phenolic acids by tailor-made DESs were systematically optimized using response surface methodology (RSM).

2. Results and Discussion

2.1. Chromatographic Conditions and Method Validation

In order to achieve a rapid and efficient analysis of the four phenolic acids (3-caffeoylquinic acid, 3,4-di-*O*-caffeoylquinic acid, 3,5-di-*O*-caffeoylquinic acid, and 4,5-di-*O*-caffeoylquinic acid) in AL, different mobile phases (including water–methanol, water–acetonitrile, formic acid water–methanol, and formic acid water–acetonitrile), flow rates (0.7 mL/min, 0.8 mL/min, and 1.0 mL/min), as well as column temperatures (15 °C, 20 °C, 25 °C, and 30 °C, Supplementary Materials Figure S1) were examined and compared. As a result, the formic acid water–acetonitrile system at 15 °C with a flow rate of 0.7 mL/min was finally selected for the suitable analysis duration, greater separation ability, and better peak shapes. The gradient elution was as follows: 0–5 min, 12% B; 5–15 min, 12–22% B; 15–25 min, 22% B; 25–35 min, 22–25% B; 35–40 min, 25–40% B. The typical HPLC chromatograms of the AL sample and four phenolic acids reference standards are shown in Figure 1.

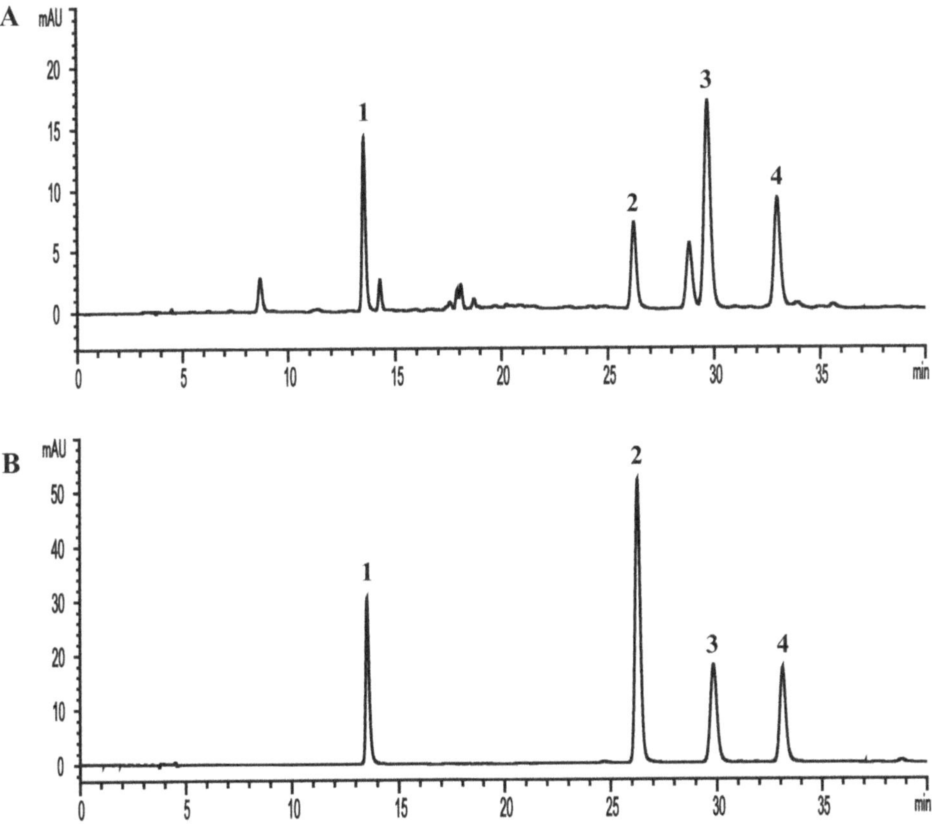

Figure 1. The typical HPLC chromatograms of (**A**) *Artemisia argyi* leaves sample (20 mg/mL) and (**B**) four phenolic acids reference standards. (1. 3-caffeoylquinic acid, 4.96 μg/mL; 2. 3,4-di-*O*-caffeoylquinic acid, 4.24 μg/mL; 3. 3,5-di-*O*-caffeoylquinic acid, 5.36 μg/mL; 4. 4,5-di-*O*-caffeoylquinic acid, 4.16 μg/mL).

Method validation of quantitative analysis was performed. The linearity, limit of detections (LODs), limit of quantifications (LOQs), precision, repeatability, stability, and accuracy for the four phenolic acids were validated. Each calibration curve was performed with six different concentrations in triplicate. All calibration curves were of good linearity with high correlation coefficient ($R^2 > 0.9997$)

over the tested range. The LODs and LOQs of the four analytes were defined by the concentration that generated peaks with signal-to-noise values of 3 and 10 using standard solutions. The precision of the developed method was determined by the intra- and interday variations. For the intraday test, the sample was analyzed six times within the same day, while for the interday test, the sample was examined in duplicates for three consecutive days. The relative standard deviations (RSDs) of intraday and interday precisions were less than 1.61% and 2.14%, respectively. For the repeatability test, six replicates of the same sample were prepared and analyzed, and for the stability test, the same sample was stored at room temperature and analyzed by replicate injection analysis at 0, 2, 4, 8, 12, and 24 h. The repeatability presented as RSDs was less than 2.71%, and the stability was less than 2.06%. The recovery was used to evaluate the accuracy of the method. Known amounts of the four phenolic acids standard solutions were added into the same samples in sextuplicate, and then extracted and analyzed with the same procedures. The recovery of each analyte was calculated by the equation: Recovery (%) = (Detected amount − Original amount)/Spiked amount × 100%. The overall recoveries of the four analytes were in the range of 101.15–102.86% with RSDs less than 1.67%. The data of method validation are shown in Supplementary Materials Table S1.

2.2. Screening of DESs for the Extraction of Phenolic Acids from AL

2.2.1. Extraction of Phenolic Acids by Binary DESs

DESs are composed of a mixture consisting of hydrogen bond acceptors (HBAs) with hydrogen bond donors (HBDs). The composition of DESs determines their physicochemical properties and consequently greatly influences extraction efficiency of natural compounds [23]. In the present study, six choline-chloride-based binary DESs, ChCl-Ma, ChCl-Ur, ChCl-Ga, ChCl-Pa, ChCl-Eg, and ChCl-Gl, were successfully synthesized (Table 1) and selected to test their extraction efficiency for phenolic acids. The high viscosity of most DESs at room temperature restricted their application due to a slow mass transfer. To overcome this problem, extraction conditions were adjusted to reduce the viscosity by increasing extraction temperature and adding a certain amount of water [24]. In the initial screening experiments, 75% DES solution in water (*v/v*) was employed, and the extraction conditions were as follows: extraction time, 30 min; extraction temperature, 50 °C; liquid–solid ratio, 50 mL/g (mL of DES/g dry weight of plant material).

Table 1. The binary DESs synthesized in this study.

NO.	Abbreviation	Component 1	Component 2	Molar Ratio
BD-1	ChCl-Ma	choline chloride	DL-malic acid	1:1
BD-2	ChCl-Ur	choline chloride	urea	1:2
BD-3	ChCl-Ga	choline chloride	glutaric acid	1:1
BD-4	ChCl-Pa	choline chloride	propanedioic acid	1:1
BD-5	ChCl-Eg	choline chloride	ethylene glycol	1:3
BD-6	ChCl-Gl	choline chloride	glycerol	1:2

Extraction yields of the four phenolic acids (3-caffeoylquinic acid, 3,4-di-*O*-caffeoylquinic acid, 3,5-di-*O*-caffeoylquinic acid, and 4,5-di-*O*-caffeoylquinic acid) with different binary DESs are shown in Figure 2. The initial screening results indicated that the extraction efficiency for phenolic acids was influenced by the types of DES solvents, and different types of DESs resulted in different extraction yields. In general, extraction yields of the four phenolic acids followed the order 3,5-di-*O*-caffeoylquinic acid > 3-caffeoylquinic acid > 4,5-di-*O*-caffeoylquinic acid > 3,4-di-*O*-caffeoylquinic acid. The phenolic acid, 3,5-di-*O*-caffeoylquinic acid, showed high extraction efficiency in ChCl-Eg with the concentration 9.35 ± 0.03 mg/g (mg/g dry weight of plant material), followed by ChCl-Ma (9.05 ± 0.05 mg/g) and ChCl-Pa (8.71 ± 0.03 mg/g). For 3-caffeoylquinic acid, ChCl-Ma (4.45 ± 0.03 mg/g) and ChCl-Ur (4.47 ± 0.03 mg/g) led to higher extraction yields. For 4,5-di-*O*-caffeoylquinic acid, ChCl-Ma (4.10 ± 0.02 mg/g), ChCl-Pa (3.90 ± 0.03 mg/g), and ChCl-Eg (3.72 ± 0.02 mg/g) exhibited higher extraction efficiency. For

3,4-di-*O*-caffeoylquinic acid, ChCl-Ur (3.32 ± 0.03 mg/g) showed higher extraction efficiency, followed by ChCl-Ma (3.03 ± 0.02 mg/g) and ChCl-Pa (2.97 ± 0.03 mg/g).In conclusion, the extraction yields of the total phenolic acids were calculated, and it was clearly shown that ChCl-Ma (BD-1) was the best binary DES for extraction of four phenolic acids from AL with the extraction yields 20.64 ± 0.08 mg/g.

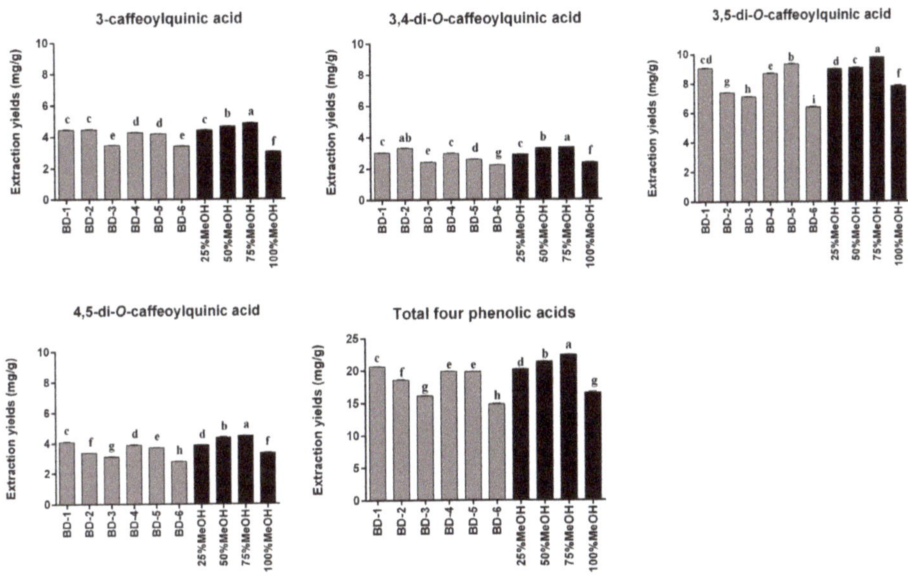

Figure 2. Extraction yields of different binary DESs and different ratios of methanol for 3-caffeoylquinic acid, 3,4-di-*O*-caffeoylquinic acid, 3,5-di-*O*-caffeoylquinic acid, 4,5-di-*O*-caffeoylquinic acid, and total four phenolic acids from *Artemisia argyi* leaves (*n* = 3). Numbers on horizontal axis are in accordance with the numbers in Table 1. Error bars indicate the SD (*n* = 3). Extraction yields which do not share the same letter are significantly different ($p < 0.05$).

In order to comprehensively compare the extraction efficiency of DESs and conventional solvents for extraction of phenolic acids from AL, different ratios of methanol (25% MeOH, 50% MeOH, 75% MeOH, and 100% MeOH), efficient solvents commonly used in the extraction of bioactive compounds from natural products, were selected as reference solvents [7]. As shown in Figure 2, among the different ratios of MeOH, 75% MeOH displayed the highest extraction efficiency for the four phenolic acids. It was clear that all the six tailor-made binary DESs exhibited higher extraction efficiency for the four phenolic acids compared with 100% MeOH (16.56 ± 0.06 mg/g). However, compared to 75% MeOH (22.41 ± 0.03 mg/g), none of the six binary DESs exhibited a higher extraction yield. Based on the results above, ChCl-Ma (BD-1) was selected as the best binary DES for extraction of phenolic acids from AL, and we attempted to synthesize a series of ternary DESs based on ChCl-Ma to enhance extraction efficiency of phenolic acids in further tests.

2.2.2. Extraction of Phenolic Acids by Ternary DESs

DESs can be synthesized from two or more components. According to the previous report, the ternary DESs forming with addition of glycerol to the binary DESs show lower melting points and viscosities [25]. Adding the third component to binary DESs may change the properties of ternary DESs and thus influence the extraction yields. In this study, based on the best binary DES (ChCl-Ma) selected above, several tailor-made ternary DESs were designed in order to further enhance the extraction yields of phenolic acids from AL. Five ternary DESs, including ChCl-Ma-Ur, ChCl-Ma-Ga, ChCl-Ma-Pa, ChCl-Ma-Eg, and ChCl-Ma-Gl, were successfully synthesized (Table 2) and further used to test their extraction efficiency for phenolic acids. The extraction yields of the four phenolic acids employing the tailor-made ternary DESs are listed in Figure 3. The results indicated that compared to the binary DESs, some tailor-made ternary DESs could change the extraction efficiency. Almost all of the tailor-made ternary DESs exhibited higher extraction efficiency for the four phenolic acids compared with the best binary DES (ChCl-Ma), except ChCl-Ma-Gl (TD-12 and TD-13). It could be speculated that the addition of the third component might reduce the viscosity of DESs, enhancing the hydrogen bond interactions between DESs and the target components, thus improving the extraction yields [26].

Table 2. The ternary DESs synthesized in this study.

NO.	Abbreviation	Component 1	Component 2	Component 3	Molar Ratio	Extraction Yield of Four Phenolic Acids (mg/g)
TD-1	ChCl-Ma-Ur	choline chloride	DL-malic acid	urea	2:1:1	21.92 ± 0.04
TD-2	ChCl-Ma-Ur	choline chloride	DL-malic acid	urea	2:2:1	22.26 ± 0.04
TD-3	ChCl-Ma-Ur	choline chloride	DL-malic acid	urea	2:1:2	22.43 ± 0.02
TD-4	ChCl-Ma-Ga	choline chloride	DL-malic acid	glutaric acid	2:1:1	21.41 ± 0.04
TD-5	ChCl-Ma-Ga	choline chloride	DL-malic acid	glutaric acid	2:2:1	21.89 ± 0.02
TD-6	ChCl-Ma-Pa	choline chloride	DL-malic acid	propanedioic acid	2:2:1	20.81 ± 0.06
TD-7	ChCl-Ma-Pa	choline chloride	DL-malic acid	propanedioic acid	2:1:1	21.30 ± 0.05
TD-8	ChCl-Ma-Pa	choline chloride	DL-malic acid	propanedioic acid	1:1:1	21.55 ± 0.04
TD-9	ChCl-Ma-Pa	choline chloride	DL-malic acid	propanedioic acid	2:1:2	22.08 ± 0.03
TD-10	ChCl-Ma-Eg	choline chloride	DL-malic acid	ethylene glycol	1:2:0.5	21.67 ± 0.04
TD-11	ChCl-Ma-Eg	choline chloride	DL-malic acid	ethylene glycol	2:2:1	22.12 ± 0.05
TD-12	ChCl-Ma-Gl	choline chloride	DL-malic acid	glycerol	1:2:0.5	20.02 ± 0.02
TD-13	ChCl-Ma-Gl	choline chloride	DL-malic acid	glycerol	2:2:1	20.09 ± 0.04

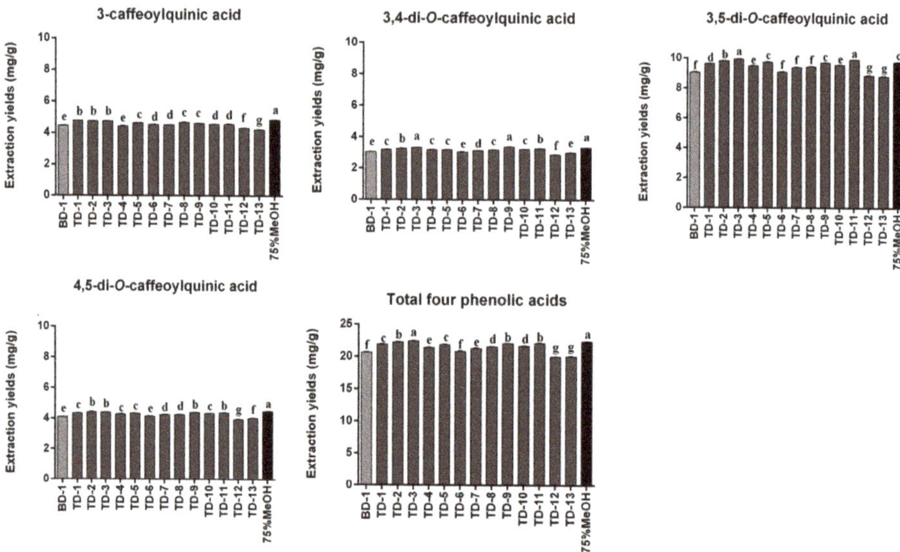

Figure 3. Extraction yields of different ternary DESs for 3-caffeoylquinic acid, 3,4-di-*O*-caffeoylquinic acid, 3,5-di-*O*-caffeoylquinic acid, 4,5-di-*O*-caffeoylquinic acid, and total four phenolic acids from *Artemisia argyi* leaves ($n = 3$). Numbers on horizontal axis are in accordance with the numbers in Tables 1 and 2. Error bars indicate the SD ($n = 3$). Extraction yields which do not share the same letter are significantly different ($p < 0.05$).

Several studies have revealed that different molar ratios of DES components result in different viscosities and surface tensions, thereby affecting the extraction efficiency of target components from natural biomass [27]. Thus, in the study, the effects of molar ratios of tailor-made ternary DESs were also investigated. As shown in Figure 3, the optimal molar ratio of ChCl-Ma-Ur was 2:1:2 (TD-3) and the extraction yields of the four phenolic acids were 22.43 ± 0.02 mg/g. The optimal molar ratio of ChCl-Ma-Ga was 2:2:1 (TD-5) and the extraction yields of the four phenolic acids were 21.89 ± 0.02 mg/g. The optimal molar ratio of ChCl-Ma-Pa was 2:1:2 (TD-9) and the extraction yields of the four phenolic acids were 22.08 ± 0.03 mg/g. The optimal molar ratio of ChCl-Ma-Eg was 2:2:1 (TD-11) and the extraction yields of the four phenolic acids were 22.12 ± 0.05 mg/g. The optimal molar ratio of ChCl-Ma-Gl was 2:2:1 (TD-13) and the extraction yields of the four phenolic acids were 20.09 ± 0.04 mg/g. None of the tailor-made ternary DESs with different molar ratios exhibited an extraction yield higher than 75% MeOH, the best conventional solvent. However, the tailor-made ternary DES, ChCl-Ma-Ur (2:1:2), produced similar extraction efficiency for the four phenolic acids compared to 75% MeOH ($p > 0.05$). Based on the results, the ChCl-Ma-Ur with a molar ratio of 2:1:2 (TD-3) was selected as the best ternary DES for extraction of phenolic acids from AL.

2.3. Optimization of the Extraction Parameters for Phenolic Acids

The above extraction investigations showcased that the tailor-made ternary DES, ChCl-Ma-Ur (2:1:2), was selected as the best ternary DES for extraction of phenolic acids from AL. In order to obtain the optimal extraction efficiency for phenolic acids from AL, several numerical variables that could affect the extraction efficiencies were optimized by RSM, a valuable statistical technique to determine the optimal values of the independent variables and to enable the user to effectively investigate the effects of multiple factors [28,29]. Similar to previous studies [30], three variables of extraction time, liquid–solid ratios, and water content in DESs were evaluated using Box–Behnken design (BBD). After determining the range of extraction factors on the basis of preliminary single-factor test, the extraction time (A), liquid–solid ratios (B), and water content (C) were varied at three levels (−1, 0, +1) as follows: A, 8.0–40.0 min; B, 17.5–57.5 mL/g; C 20–70%. The total extraction amounts of the four phenolic acids were taken as the responses of the design experiments. The experimental orders, levels of variables, and response values are summarized in Table 3.

Table 3. The experimental orders, levels of variables, and response values in Box–Behnken design.

Run	Factors			Responses
	Extraction Time (A, min)	Liquid–Solid Ratios (B, mL/g)	Water Content (C, %)	Total Extraction Amounts of Four Phenolic Acids (mg/g)
1	24.0	37.5	45	22.60
2	40.0	57.5	45	22.61
3	24.0	17.5	20	18.47
4	8.0	37.5	20	16.13
5	24.0	37.5	45	22.10
6	40.0	17.5	45	21.96
7	24.0	57.5	70	20.85
8	24.0	37.5	45	21.40
9	8.0	57.5	45	22.34
10	24.0	17.5	70	18.86
11	24.0	57.5	20	15.51
12	24.0	37.5	45	22.65
13	8.0	17.5	45	21.27
14	40.0	37.5	70	19.84
15	24.0	37.5	45	22.02
16	8.0	37.5	70	20.06
17	40.0	37.5	20	17.00

Experiments conducted according to the design resulted in a second-order polynomial equation for total extraction amounts (Y) expressed using coded variables (A, B, and C) as follows:

$$Y = 22.16 - 0.20A + 0.095B + 1.56C - 0.11AB - 0.27AC + 1.24BC - 0.14A^2 + 0.027B^2 - 3.76C^2$$

The model was evaluated in terms of the square of correlation coefficient (R^2) and the lack of fit by the analysis of variance (ANOVA) at the 95% confidence level (Table 4). The resulting R^2 value was 0.9765, indicating that the experimental data were in relatively good agreement with predicted extraction yields. The lack-of-fit value, which evaluates the failure of the model to represent the data in the experimental domain points, was insignificant for the response with p-value of 0.3789 ($p > 0.05$).

Table 4. The ANOVA results of the quadratic multiple regression model for phenolic acids.

Source	Sum of Squares	df	Mean Square	F Value	p-Value Prob > F	Significance
Model	86.54	9	9.62	32.38	<0.0001	significant
A	0.32	1	0.32	1.08	0.3323	
B	0.073	1	0.073	0.24	0.6359	
C	19.53	1	19.53	65.78	<0.0001	
AB	0.045	1	0.045	0.15	0.7085	
AC	0.3	1	0.3	1.02	0.3468	
BC	6.13	1	6.13	20.64	0.0027	
A^2	0.08	1	0.08	0.27	0.6188	
B^2	3.10×10^{-3}	1	3.10×10^{-3}	0.01	0.9215	
C^2	59.51	1	59.51	200.42	<0.0001	
Residual	2.08	7	0.3			
Lack of Fit	1.04	3	0.35	1.34	0.3789	not significant
Pure Error	1.04	4	0.26			
R^2	0.9765					

Statistical analysis and 3D response plots (Figure 4) illustrated the significant variables affecting extraction yields of phenolic acids and the interaction effects between the variables. In the model, the water content (C) showed significant effects on the extraction efficiency of phenolic acids ($p < 0.0001$). Based on the adequate model, the calculated optimum conditions for the extraction of phenolic acids from AL were as follows: extraction time, 23.5 min; liquid–solid ratio, 57.5 mL/g; water content, 54%. Triplicate experiments were carried out under the optimal extraction conditions and mean values of experimental results were compared with the predicted values. Under the optimum conditions, the total extraction amounts of the four phenolic acids were 22.80 mg/g, which were closed to the predicted values of 22.79 mg/g. The results obtained through confirmation experiments indicated the model is adequate for predicting the expected optimization.

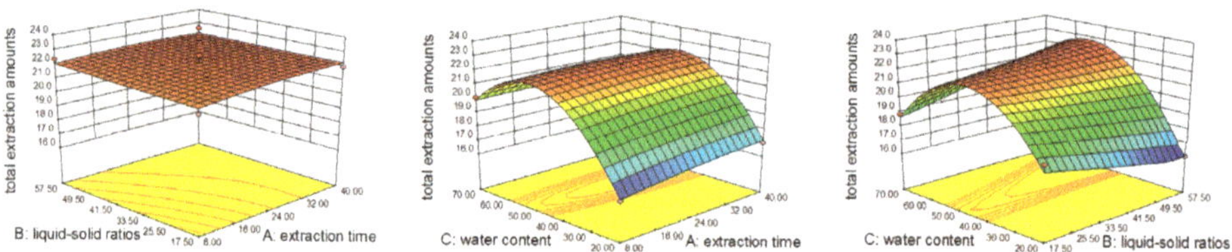

Figure 4. Response surface plots of the model for extraction of phenolic acids from *Artemisia argyi* leaves. (A. extraction time, min; B. liquid–solid ratios, mL/g; C. water content, %).

3. Materials and Methods

3.1. Materials and Reagents

AL samples were purchased from a Chinese herbal medicine market (Anguo, China). The samples were dried in the shade and stored in the desiccator. The AL samples were authenticated by Prof. Yuguang Zheng from Department of Pharmacognosy, Hebei University of Chinese Medicine and the voucher specimens were deposited in Hebei University of Chinese Medicine, Shijiazhuang, China.

Four phenolic acids reference compounds, 3-caffeoylquinic acid, 3,4-di-*O*-caffeoylquinic acid, 3,5-di-*O*-caffeoylquinic acid, and 4,5-di-*O*-caffeoylquinic acid, were purchased from Chengdu Must Bio-technology Co., Ltd. (Chengdu, China). The purities of the four reference compounds were determined to be higher than 98% by high-performance liquid chromatography diode array detection analysis. The structures of the four phenolic acids are shown in Figure 5.

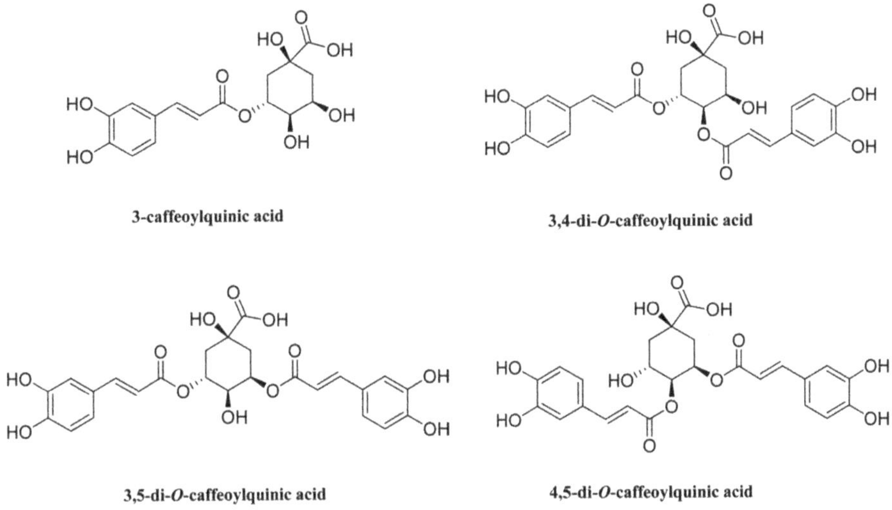

3-caffeoylquinic acid

3,4-di-*O*-caffeoylquinic acid

3,5-di-*O*-caffeoylquinic acid

4,5-di-*O*-caffeoylquinic acid

Figure 5. Chemical structures of 3-caffeoylquinic acid, 3,4-di-*O*-caffeoylquinic acid, 3,5-di-*O*-caffeoylquinic acid, and 4,5-di-*O*-caffeoylquinic acid.

Chemical compounds for DES preparation including choline chloride (ChCl), DL-malic acid (Ma), urea (Ur), glutaric acid (Ga), propanedioic acid (Pa), ethylene glycol (Eg), glycerol (Gl) were obtained from Aladdin Reagent Company (Shanghai, China). Acetonitrile, methanol, and formic acid (chromatographic grade) were purchased from Merck (Darmstadt, Germany). Deionized water was prepared by a Milli-Q water purification system (Millipore, Billerica, MA, USA). Other reagents and chemicals used in this work were of analytical grade.

3.2. HPLC Analysis

HPLC analysis was performed on an Agilent 1260 HPLC system equipped with a quaternary pump, a degasser, an autosampler, a thermostated column compartment, and a diode array detector (Agilent Technologies, Palo Alto, CA, USA). Chromatographic separation was achieved on an Agilent SB C18 column (4.6 × 250 mm, 5 µm). The mobile phase was composed of 0.1% formic acid water (A) and acetonitrile (B) with a gradient elution as follows: 0–5 min, 12% B; 5–15 min, 12–22% B; 15–25 min, 22% B; 25–35 min, 22–25% B; 35–40 min, 25–40% B. The detection wavelength was 330 nm. The flow rate was set at 0.7 mL/min and the column temperature was set at 15 °C.

3.3. Preparation of DESs

The DESs were synthesized according to the previous studies [31]. Briefly, different DESs were obtained by simply mixing hydrogen bond acceptors (HBAs) and hydrogen bond donors (HBDs) together in a proper molar ratio with constant stirring at 80 °C until a clear and homogeneous liquid formed.

3.4. Extraction of Phenolic Acids From AL

The AL samples were powdered and screened through 40 mesh sieves. An accurately weighed powder (20 mg) was extracted with 1 mL of different solvents (different DESs, 25% MeOH, 50% MeOH, 75% MeOH, and 100% MeOH) in a 2 mL centrifuge tube by ultrasonic cleaner (KQ5200B, 200W, 40 kHz, Kunshan, China) for 30 min. After extraction, the extracted solution was centrifuged at 13,000 rpm/min for 10 min. Then, 0.1 mL of supernatant was sampled and diluted 10-fold with methanol. The diluted solution (2 µL) was injected into the HPLC instrument for analysis. Three replicates of each sample were prepared and analyzed ($n = 3$).

3.5. DESs Tailoring

In order to determine the effect of the DES compositions on the extraction of the four phenolic acids from AL, six ChCl-based binary DESs, including ChCl-Ma, ChCl-Ur, ChCl-Ga, ChCl-Pa, ChCl-Eg, and ChCl-Gl, were successfully synthesized in initial screening. Then, the extraction efficiency of the different binary DESs for the four phenolic acids was investigated and the optimal binary DES was selected. Next, different efficient components were added into the optimal binary DES obtained in initial screening to design different ternary DESs. Five tailored ternary DESs with different molar ratios, including ChCl-Ma-Ur, ChCl-Ma-Ga, ChCl-Ma-Pa, ChCl-Ma-Eg, and ChCl-Ma-Gl, were synthesized and the extraction efficiencies for the four phenolic acid compounds were tested.

3.6. Optimization of the DES Extraction Parameters for Phenolic Acids

The optimal extraction parameters for the phenolic acids from AL were obtained using RSM. RSM was performed using the Design-Expert Ver. 8.0.6 (Stat-Ease Inc., Minneapolis, MN, USA). After determining the range of extraction variables on the basis of preliminary single-factor test, Box–Behnken design (BBD) was used to find the optimal values for three independent variables: extraction time (A), liquid–solid ratios (B), and water content (%) in DESs (C). The total extraction amounts of the four phenolic acids were taken as the response of the design experiments. Regression analysis was performed according to the experimental data. Subsequently, additional confirmation experiments were conducted to confirm the validity of the statistical experimental strategies.

4. Conclusions

In this work, a green and efficient extraction method using tailor-made DESs as extraction solvents was established for extraction of bioactive phenolic acids from AL. Six binary DESs and five ternary DESs were successfully synthesized and used to extract four phenolic acids from AL. The results indicated that the tailor-made DESs were efficient solvents for the extraction of phenolic acids. The ternary DES, ChCl-Ma-Ur with a molar ratio of 2:1:2, proved to be the most efficient solvent for phenolic acid extraction from AL. Moreover, the optimal extraction conditions for phenolic acids by ChCl-Ma-Ur (2:1:2) were determined using RSM and the optimal extraction conditions are extraction time, 23.5 min; liquid–solid ratio, 57.5 mL/g; and water content, 54%. The present study suggests that the DES-based extraction method is efficient and sustainable for extraction of bioactive phenolic acids from AL. The DESs are truly designed and efficient solvents which could be used as green solvents for the extraction of bioactive compounds from natural products and have a great potential as possible alternatives to those organic solvents in health-related areas such as food and pharmaceuticals.

Supplementary Materials: The Supplementary Materials are available online.

Author Contributions: Conceptualization, L.D. and L.G.; Data curation, C.Z. (Chenmeng Zhang); Investigation, L.D. and C.Z. (Chenmeng Zhang); Methodology, C.Z. (Chenjing Zhang); Project administration, Y.Z.; Resources, Z.X.; Writing—original draft, L.D. and L.G.; Writing—review & editing, Y.Z. and L.G.

Funding: This research was funded by National Natural Science Foundation of China (81803697, 81803685), Natural Science Foundation of Hebei Province (H2018205185, H2018423032), Research Foundation of Hebei Province Education Department (BJ2019006, QN2017093), Research Foundation of Hebei Provincial Administration of Traditional Chinese Medicine (2018108), and Innovation Team of Hebei Province Modern Agricultural Industry Technology System (HBCT2018060205).

Conflicts of Interest: The authors declare no conflict of interest.

References

1. Abad, M.J.; Bedoya, L.M.; Apaza, L.; Bermejo, P. The *artemisia*, Genus: A review of bioactive essential oils. *Molecules* **2012**, *17*, 2542–2566. [CrossRef] [PubMed]
2. Bora, K.S.; Sharma, A. The genus *Artemisia*: A comprehensive review. *Pharm. Biol.* **2011**, *49*, 101–109. [CrossRef] [PubMed]
3. Guan, X.; Ge, D.; Li, S.; Huang, K.; Liu, J.; Li, F. Chemical composition and antimicrobial activities of *Artemisia argyi* Lévl. et Vant essential oils extracted by simultaneous distillation-extraction, subcritical extraction and hydrodistillation. *Molecules* **2019**, *24*, 483. [CrossRef] [PubMed]
4. Li, S.; Zhou, S.; Yang, W.; Meng, D. Gastro-protective effect of edible plant *Artemisia argyi* in ethanol-induced rats via normalizing inflammatory responses and oxidative stress. *J. Ethnopharmacol.* **2018**, *214*, 207–217. [CrossRef] [PubMed]
5. Han, B.; Xin, Z.; Ma, S.; Liu, W.; Zhang, B.; Ran, L.; Yi, L.; Ren, D. Comprehensive characterization and identification of antioxidants in Folium Artemisiae Argyi using high-resolution tandem mass spectrometry. *J. Chromatogr. B* **2017**, *1063*, 84–92. [CrossRef] [PubMed]
6. Li, N.; Mao, Y.; Deng, C.; Zhang, X. Separation and identification of volatile constituents in *Artemisia argyi* flowers by GC-MS with SPME and steam distillation. *J. Chromatogr. Sci.* **2008**, *46*, 401–405. [CrossRef] [PubMed]
7. Guo, L.; Zhang, D.; Xue, Z.J.; Jiao, Q.; Liu, A.P.; Zheng, Y.G.; Liu, E.H.; Duan, L. Comparison of Artemisiae argyi Folium and Artemisiae lavandulaefoliae Folium by simultaneous determination of multi-components with single reference standard method and chemometric analysis. *Phytochem. Anal.* **2019**, *30*, 14–25. [CrossRef] [PubMed]
8. Kim, K.O.; Lee, D.; Hiep, N.T.; Song, J.H.; Lee, H.J.; Lee, D.; Kang, K.S. Protective effect of phenolic compounds isolated from Mugwort (*Artemisia argyi*) against contrast-induced apoptosis in kidney epithelium cell line LLC-PK1. *Molecules* **2019**, *24*, 195. [CrossRef]
9. Lee, D.; Kim, C.E.; Park, S.Y.; Kim, K.O.; Hiep, N.T.; Lee, D.; Jang, H.J.; Lee, J.W.; Kang, K.S. Protective effect of Artemisia argyi and its flavonoid constituents against contrast-induced cytotoxicity by iodixanol in LLC-PK1 cells. *Int. J. Mol. Sci.* **2018**, *19*, 1387. [CrossRef]
10. Belwal, T.; Ezzat, S.M.; Rastrelli, L.; Bhatt, I.D.; Daglia, M.; Baldi, A.; Devkota, H.P.; Orhan, I.E.; Patra, J.K.; Das, G.; et al. A critical analysis of extraction techniques used for botanicals: Trends, priorities, industrial uses and optimization strategies. *Trends Analyt. Chem.* **2018**, *100*, 82–102. [CrossRef]
11. Chemat, F.; Vian, M.A.; Cravotto, G. Green extraction of natural products: Concept and principles. *Int. J. Mol. Sci.* **2012**, *13*, 8615–8627. [CrossRef] [PubMed]
12. Abbott, A.P.; Capper, G.; Davies, D.L.; Rasheed, R.K.; Tambyrajah, V. Novel solvent properties of choline chloride/urea mixtures. *Chem. Commun.* **2003**, *1*, 70–71. [CrossRef]
13. Zhang, Q.; De Oliveira, K.; Royer, S.; Jérôme, F. Deep eutectic solvents: Syntheses, properties and applications. *Chem. Soc. Rev.* **2012**, *41*, 7108–7146. [CrossRef] [PubMed]
14. Huang, J.; Guo, X.; Xu, T.; Fan, L.; Zhou, X.; Wu, S. Ionic deep eutectic solvents for the extraction and separation of natural products. *J. Chromatogr. A* **2019**. [CrossRef] [PubMed]
15. Zhao, H. DNA stability in ionic liquids and deep eutectic solvents. *J. Chem. Technol. Biotechnol.* **2015**, *90*, 19–25. [CrossRef]
16. Fernández, M.L.Á.; Boiteux, J.; Espino, M.; Gomez, F.J.V.; Silva, M.F. Natural deep eutectic solvents-mediated extractions: The way forward for sustainable analytical developments. *Anal. Chim. Acta* **2018**, *1038*, 1–10.
17. Cunha, S.C.; Fernandes, J.O. Extraction techniques with deep eutectic solvents. *TrAC Trend Anal. Chem.* **2018**, *105*, 225–239. [CrossRef]
18. Zainal-Abidin, M.H.; Hayyan, M.; Hayyan, A.; Jayakumar, N.S. New horizons in the extraction of bioactive compounds using deep eutectic solvents: A review. *Anal. Chim. Acta* **2017**, *979*, 1–23. [CrossRef]
19. Ruesgas-Ramón, M.; Figueroa-Espinoza, M.C.; Durand, E. Application of deep eutectic solvents (DES) for phenolic compounds extraction: Overview, challenges, and opportunities. *J. Agric. Food. Chem.* **2017**, *65*, 3591–3601. [CrossRef]
20. Jablonský, M.; Škulcová, A.; Malvis, A.; Šima, J. Environmentally-friendly extraction of flavonoids from *Cyclocarya paliurus* (Batal.) Iljinskaja leaves with deep eutectic solvents and evaluation of their antioxidant activities. *Molecules* **2018**, *23*, E2110.

21. Duan, L.; Dou, L.L.; Guo, L.; Li, P.; Liu, E. Comprehensive evaluation of deep eutectic solvents in extraction of bioactive natural products. *ACS Sustainable Chem. Eng.* **2016**, *4*, 2405–2411. [CrossRef]

22. Jeong, K.M.; Lee, M.S.; Nam, M.W.; Zhao, J.; Jin, Y.; Lee, D.K.; Kwon, S.W.; Jeong, J.H.; Lee, J. Tailoring and recycling of deep eutectic solvents as sustainable and efficient extraction media. *J. Chromatogr. A* **2015**, *1424*, 10–17. [CrossRef] [PubMed]

23. Li, X.; Row, K.H. Development of deep eutectic solvents applied in extraction and separation. *J. Sep. Sci.* **2016**, *39*, 3505–3520. [CrossRef] [PubMed]

24. Dai, Y.; Row, K.H. Application of natural deep eutectic solvents in the extraction of quercetin from vegetables. *Molecules* **2019**, *24*, 2300. [CrossRef] [PubMed]

25. Maugeri, Z.; de Maria, P.D. Novel choline-chloride-based deep-eutectic-solvents with renewable hydrogen bond donors: Levulinic acid and sugar-based polyols. *RSC Adv.* **2012**, *2*, 421–425. [CrossRef]

26. Xu, M.; Ran, L.; Chen, N.; Fan, X.; Ren, D.; Yi, L. Polarity-dependent extraction of flavonoids from citrus peel waste using a tailor-made deep eutectic solvent. *Food Chem.* **2019**, *297*, 124907. [CrossRef] [PubMed]

27. Cao, J.; Wang, H.; Zhang, W.; Cao, F.; Ma, G.; Su, E. Tailor-made deep eutectic solvents for simultaneous extraction of five aromatic acids from *Ginkgo biloba* leaves. *Molecules* **2018**, *23*, 3214. [CrossRef] [PubMed]

28. Guo, N.; Ping, K.; Jiang, Y.W.; Wang, L.T.; Niu, L.J.; Liu, Z.M.; Fu, Y.J. Natural deep eutectic solvents couple with integrative extraction technique as an effective approach for mulberry anthocyanin extraction. *Food Chem.* **2019**, *296*, 78–85. [CrossRef] [PubMed]

29. Ma, Y.; Liu, M.; Tan, T.; Yan, A.; Guo, L.; Jiang, K.; Tan, C.; Wan, Y. Deep eutectic solvents used as extraction solvent for the determination of flavonoids from *Camellia oleifera* flowers by high-performance liquid chromatography. *Phytochem. Anal.* **2018**, *29*, 639–648. [CrossRef]

30. Wang, H.; Ma, X.; Cheng, Q.; Wang, L.; Zhang, L. Deep eutectic solvent-based ultrahigh pressure extraction of baicalin from *Scutellaria baicalensis* Georgi. *Molecules* **2018**, *23*, 3233. [CrossRef]

31. Duan, L.; Zhang, W.H.; Zhang, Z.H.; Liu, E.H.; Guo, L. Evaluation of natural deep eutectic solvents for the extraction of bioactive flavone C-glycosides from flos trollii. *Microchem. J.* **2018**, *145*, 180–186. [CrossRef]

Sample Availability: Samples of *Artemisia argyi* leavesare available from the authors.

Article

Biochemical, Micronutrient and Physicochemical Properties of the Dried Red Seaweeds *Gracilaria edulis* and *Gracilaria corticata*

Thomas Rosemary [1], Abimannan Arulkumar [1,2], Sadayan Paramasivam [1,*], Alicia Mondragon-Portocarrero [3] and Jose Manuel Miranda [3]

[1] Department of Oceanography and Coastal Area Studies, School of Marine Sciences, Alagappa University, Thondi Campus, Thondi-623 409, India; rosemarythomas@gmail.com (T.R.); aruul3@gmail.com (A.A.)

[2] Department of Biotechnology, Achariya Arts and Science College, Villianur, Puducherry-60 110, India

[3] Departamento de Química Analítica, Nutrición y Bromatología, Facultade de Veterinaria, Universidade de Santiago de Compostela, Pabellón 4, Planta Baja, 27002 Lugo, Spain; aliciamondragon@yahoo.com (A.M.-P.); josemanuel.miranda@usc.es (J.M.M.)

* Correspondence: drparamsan@gmail.com; Tel.: +91-9047733300; Fax: +91-4565-225525

Academic Editor: Gavino Sanna
Received: 25 April 2019; Accepted: 11 June 2019; Published: 14 June 2019

Abstract: The present study sought to evaluate the nutritional composition and physicochemical properties of two dried commercially interesting edible red seaweeds, *Gracilaria corticata* and *G. edulis*. Proximate composition of the dried seaweeds revealed a higher content in carbohydrates (8.30 g/100 g), total crude protein (22.84 g/100 g) and lipid content (7.07 g/100 g) in *G. corticata* than in *G. edulis*. Fatty acids profile showed that *G. corticata* samples contain higher concentrations of saturated fatty acids, such as palmitic and stearic acids, and polyunsaturated ones such as α-linolenic and docosahexaenoic acids. Contrariwise, *G. edulis* contained higher amounts of monounsaturated oleic acid. Total amino acid content was 76.60 mg/g in *G. corticata* and 65.42 mg/g in *G. edulis*, being the essential amino acid content higher in *G. edulis* (35.55 mg/g) than in *G. corticata* (22.76 mg/g). Chlorophyll *a* was found in significantly higher amounts in *G. edulis* (17.14 µg/g) than *G. corticata*, whereas carotenoid content was significantly higher in *G. corticata* (12.98 µg/g) than in *G. edulis*. With respect to physical properties, both water- and oil-holding capacities were similar in both seaweeds, whereas swelling capacity was higher in *G. edulis*. In view of the results, the present study suggests that *G. corticata* and *G. edulis* contains important nutrients for human health and are possible natural functional foods.

Keywords: water holding capacity; seaweeds; *Gracilaria*; carbohydrates; fatty acids; vitamins

1. Introduction

Seaweeds are very important natural resources from the oceans that are employed as human foods and animal feeds in their whole form, and as sources of polysaccharides (mainly alginates, carrageenans and agar), carotenoids, lipids, vitamins, minerals, dietary fiber, proline and amino acids for use in food and pharmaceutical industry [1]. Seaweeds have been included for a long time in the traditional diet of East Asian countries such as Japan, Korea and China; more recently, their presence in all forms in the diet of Western countries has been progressively increasing [2].

Seaweeds are considered healthy foods because, despite their low caloric content, they are rich in important nutrients such as protein, essential amino acids, vitamins, minerals and some bioactive compounds [1]. Seaweeds are also an excellent source of both soluble and insoluble dietary fiber. Among red algae, the genus *Gracilaria* contains a broad diversity of valuable contents for human nutrition and are one of the world's most cultivated and valuable marine seaweed [3]. Its lipid content is low (1–5% dry weight, DW) [1], but it contains docosahexaenoic acid (DHA) which is recognized as the most important *n*-3 polyunsaturated fatty acid (PUFA) to reduce the risk of cardiovascular diseases [4,5]. In particular, *n*-3 PUFAs act as excellent antioxidants, strengthening the cell membrane, repairing damaged cells and tissues, improving heart function and fighting against cancer [6]. *n*-3 PUFAs were also found to prevent the growth of atherosclerotic plaque that affects blood clotting and blood pressure and improve the immune function, while *n*-6 PUFAs decrease low-density lipoprotein cholesterol and may also decrease high-density lipoprotein, cholesterol which reduces heart disease risk [7].

With respect to their protein content, the most abundant amino acids in *Gracilaria* species are aspartic acid, alanine, glutamic acid and glutamine. These amino acids provide the typical flavor of algae and accumulate in response to stress conditions [8]. *Gracilaria* is also a good source of both soluble and insoluble dietary fiber, so it can be employed as a potential alternative to cereal-based fiber in Western countries [1]. Soluble dietary fiber helps to increase viscosity and reduce glycemic response and plasma cholesterol in humans [1]. Insoluble dietary fiber improves the bulking effect caused by water absorption in feces and thus contributes to weight management, improvement of cardiovascular and gastrointestinal functions and cancer prevention [1]. Polysaccharides isolated from red seaweeds show potent antibacterial, antiviral, antioxidant, anticoagulant and anti-inflammatory activity [9].

Seaweeds such as *Gracilaria* can concentrate minerals from seawater and reach a mineral content 10–20 times higher than that of terrestrial plants [10]. Consequently, they are a valuable source of minerals, with important human nutrition functions [11,12]. Chlorophyll, an important pigment constituent present in algae, has positive effects on inflammation, oxidation and wound healing [13]. Chlorophyll acts directly as a reducer of free radicals and has the potential to protect lymphocytes against oxidative DNA damage by free radicals [14]. Moreover, a large number of potentially bioactive compounds such as phenols, polyphenols, terpenes, steroids, halogenated ketones and alkanes, fucoxanthin, polyphloroglucinol and bromophenols have been isolated [15–17].

However, the nutrient profile of seaweeds such as *Gracilaria* is influenced by different factors such as seaweed species, habitat, maturity stage, season, water temperature and the sampling conditions and method employed in the determinations [1,2]. *Gracilaria edulis* and *G. corticata* is abundantly available in almost all seasons in Palk Bay, on the southeast coast of India, rather than other *Gracilaria* sp. Both *G. edulis* and *G. corticata* are commercially important and commonly edible seaweeds in India. These two algae exhibited in a previous work high biological activities (proximate composition, antioxidant, antibacterial, and biopreservative effects in seafoods during preservation and extended shelf life than other *Gracilaria* species in a previous work [18]. Thus, the present study sought to evaluate and compare the chemical composition (proximate composition, lipid profile, amino acids, vitamins and pigments such as chlorophyll and carotenoids) and physicochemical properties of both, *Gracilaria corticata* and *Gracilaria edulis* from the Thondi coast of Palk Bay, southeast India.

2. Results

Proximate, polysaccharide content and fatty acids profile of both *G. corticata* and *G. edulis* in a DW basis are shown in Table 1.

Table 1. Proximate composition (g/100 g dry weight seaweed) *G. corticata* and *G. edulis*.

Parameter	*G. corticata*	*G. edulis*
Moisture	8.40 ± 0.65 [b]	10.40 ± 0.69 [a]
Protein	22.84 ± 0.87 [b]	25.29 ± 0.67 [a]
Fat	7.07 ± 0.33 [a]	4.76 ± 0.73 [b]
Carbohydrates	8.30 ± 1.89 [a]	4.71 ± 0.60 [b]
Ash	8.10 ± 0.49	7.36 ± 0.39
Polysaccharides	49.64 ± 3.89 [a]	38.02 ± 4.32 [b]

Values are mean ± standard deviation, $n = 3$. [a–b] values with different superscripts within the same line were significantly different.

The crude polysaccharide content found for *G. corticata* and *G. edulis* was 49.64 g/100 g and 38.02 g/100 g, respectively. The moisture content (in dried seaweeds) of *G. corticata* and *G. edulis* was 8.40 g/100 g and 10.40 g/100 g, respectively. With respect to proximate composition, important differences were obtained for the two seaweeds investigated. Carbohydrates and fat content were significantly higher in *G. corticata*, whereas protein content was significantly higher in the case of *G. edulis*.

With respect to fatty acids profile, total fatty acid content, expressed as g fatty acids methyl esters (FAME)/100 g total fat, of *G. corticata* and *G. edulis* was 5.49 ± 0.30 g/100 g and 3.92 ± 0.13 g/100 g, respectively (Table 2). The main saturated fatty acids (SFAs) found in both *G. corticata* and *G. edulis* were palmitic acid (C16:0), margaric acid (C17:0) and stearic acid (C18:0). With respect to PUFAs, linoleic acid (C18:2n-6), α-linolenic acid (C18:3n-3), stearidonic acid (C18:4n-3) and DHA (C22:6n-3) were found in both seaweeds. In the case of monounsaturated fatty acids (MUFAs), only oleic acid (C18:1) was detected in relevant amounts in both *G. corticata* and *G. edulis*. Margaric, linoleic and stearidonic acids were found in similar amounts in *G. corticata* and *G. edulis*. Palmitic, stearic, α-linolenic acid were found in higher amounts in *G. corticata* than in *G. edulis*. Overall, in G. corticata, SFAs accounted 49.4% of total fatty acids, MUFAs accounted a 3.3% and PUFAs accounted a 47.3%, whereas in the case of *G. edulis*, SFAs accounted 43.9% of total fatty acids, MUFAs accounted a 27% of total fatty acids, and PUFAs accounted a 29%.

Table 2. Fatty acids profile (g fatty acid methyl esters/100 g total fat) of *G. corticata* and *G. edulis*.

Parameter	*G. corticata*	*G. edulis*
Palmitic acid	1.22 ± 0.04 [a]	0.63 ± 0.09 [b]
Margaric acid	0.16 ± 0.25	0.15 ± 0.10
Stearic acid	1.31 ± 0.03 [a]	0.93 ± 0.05 [b]
Oleic acid	0.18 ± 0.12 [b]	1.05 ± 0.05 [a]
Linoleic acid	0.63 ± 0.04	0.65 ± 0.18
α-Linolenic acid	1.26 ± 0.04 [a]	0.14 ± 0.04 [b]
Stearidonic acid	0.21 ± 0.01	0.22 ± 0.05
Docosohexaenoic acid	0.48 ± 0.14 [a]	0.12 ± 0.18 [b]
$\sum$FA	5.49 ± 0.30 [a]	3.92 ± 0.13 [b]

$\sum$FA = Total fatty acids. Values are mean ± standard deviation, $n = 3$. [a–b] values with different superscripts within the same line were significantly different.

The protein of *G. corticata* and *G. edulis* is shown in Table 3. The total amino acid content was higher in *G. corticata* (76.60 ± 5.14 mg/g), than in *G. edulis* (65.42 ± 3.58 mg/g). These values are comparable to their corresponding crude protein content of 22.84 ± 0.87 and 25.29 ± 0.67 g/100 g, respectively, indicating that the amount of non-protein nitrogenous materials in these red seaweeds is low.

Nine essential amino acids (EAAs), and 11 non-essential amino acids (NEAAs), were found in both *G. corticata* and *G. edulis*. Total EAAs where significantly higher in *G. edulis* (35.55 ± 1.75 mg/g) than in *G. corticata* (22.76 ± 1.81 mg/g), whereas total NEAAs where higher in *G. corticata* (36.14 ± 3.33 mg/g) than in *G. edulis* (29.86 ± 1.83 mg/g). The EAAs/total amino acid ratio suggests that more than 50% of the amino acids found in *G. edulis* are EAAs. The results also indicate a good ratio of essential amino acids to non-essential amino acids in *G. corticata* (0.62 ± 0.54 mg/g) and *G. edulis* (1.19 ± 0.95 mg/g). It was noted that a much higher concentration of the essential amino acid threonine (20.57 ± 0.62 mg/g) was found in *G. edulis* than in *G. corticata*. Contrariwise, alanine content was much higher in *G. corticata* (21.11 ± 0.54 mg/g) than in *G. edulis* (1.46 ± 0.18 mg/g). Aspartic acid content was similar in both seaweeds.

The mineral content of *G. corticata* and *G. edulis* is shown in Table 4. *G. corticata* showed a higher content of Mg (463.23 ± 8.87 mg/kg) and Fe (1072.48 ± 20.97 mg/kg) than *G. edulis*. Moreover, *G. edulis* was found to possess more of trace elements like Zn (42.73 ± 2.12 mg/kg) and Cu (14.61 ± 0.46 mg/kg) than *G. corticata*. In view of the present results, both *G. corticata* and *G. edulis* contain an adequate amount of minerals, which suggests that these seaweeds could act as important sources of mineral supplements which are essential for human nutrition.

Table 3. Protein composition of *G. corticata* and *G. edulis*.

Amino Acid	*G. corticata*	*G. edulis*
Aspartic acid	14.37 ± 0.78	12.67 ± 0.64
Glutamic acid	2.54 ± 0.06	2.77 ± 0.15
Asparagine	1.45 ± 0.05 [b]	1.89 ± 0.15 [a]
Serine	2.23 ± 0.18 [b]	2.73 ± 0.13 [a]
Glutamine	2.01 ± 0.7 [b]	2.42 ± 0.29 [a]
Glycine	4.71 ± 0.18 [a]	3.42 ± 0.27 [b]
Theronine *	1.32 ± 0.09 [b]	20.57 ± 0.62 [a]
Arginine *	3.41 ± 0.30	3.33 ± 0.17
Alanine	21.11 ± 0.54 [a]	1.46 ± 0.18 [b]
Cysteine	1.49 ± 0.30	1.27 ± 0.06
Tyrosine *	1.25 ± 0.15 [b]	2.50 ± 0.24 [b]
Histidine	2.46 ± 0.27 [a]	0.18 ± 0.02 [a]
Valine *	0.16 ± 0.01	0.15 ± 0.02
Methionine *	8.73 ± 0.31 [a]	4.98 ± 0.48 [b]
Isoleucine *	2.53 ± 0.16 [a]	1.22 ± 0.07 [b]
Phenylalanine *	1.42 ± 0.17 [b]	2.20 ± 0.10 [a]
Leucine *	1.58 ± 0.35 [a]	0.38 ± 0.02 [b]
Lysine *	2.37 ± 0.27 [a]	0.22 ± 0.03 [b]
Proline	0.47 ± 0.20	0.46 ± 0.18
Tryptophan	1.00 ± 0.07 [a]	0.59 ± 0.16 [b]
Total amino acids	76.60 ± 5.14 [a]	65.42 ± 3.58 [b]
Total EAAs	22.76 ± 1.81 [b]	35.55 ± 1.75 [a]
Total NEAAs	36.14 ± 3.33 [a]	29.86 ± 1.83 [b]
EAAs/Total AAs	0.29 ± 0.35 [b]	0.54 ± 0.48 [a]
EAAs/NEAAs	0.62 ± 0.54 [b]	1.19 ± 0.95 [a]

Values are mean ± standard deviation, *n* = 3 expressed as mg/g seaweed in a dry weight basis. * EAAs: Essential amino acids; NEAAs: Non-essential amino acids. [a–b] values with different superscripts within the same line are significantly different.

Table 4. Mineral content (mg/kg) in *G. corticata* and *G. edulis*.

Minerals	*G. corticata*	*G. edulis*
Zn	31.52 ± 0.69 [b]	42.73 ± 2.12 [a]
Cu	11.42 ± 0.72 [b]	14.61 ± 0.46 [a]
Mg	463.23 ± 8.87 [a]	89.56 ± 0.77 [b]
Fe	1072.48 ± 20.97 [a]	557.36 ± 0.57 [b]

Values are mean ± standard deviation, *n* = 3 on dry weight basis. Macro minerals: Zinc (Zn), copper (Cu), magnesium (Mg) and iron (Fe). [a–b] values with different superscripts within the same line are significantly different.

For both *G. corticata* and *G. edulis*, the presence of water-soluble vitamins (vitamin B_1, vitamin B_2, vitamin B_3, vitamin B_6, vitamin B_9 and vitamin C) and fat-soluble vitamins (vitamin A and vitamin E) was found, as shown in Table 5. *G. corticata* had a higher vitamin A (2.67 ± 0.31 mg/g vs. 2.14 ± 0.17 mg/g) and vitamin B_9 contents (1.00 ± 0.07 mg/g vs. 0.45 ± 0.06 mg/g) than *G. edulis*, whereas *G. edulis* showed a significantly higher content of vitamin B_2 (1.54 ± 0.39 mg/g vs. 0.05 ± 0.01 mg/g) and vitamin B_6 (4.77 ± 0.23 mg/g vs. 3.79 ± 0.30 mg/g) than *G. corticata*.

Table 5. Vitamin composition (mg/g), chlorophyll and carotenoid content (µg/g) of *G. corticata* and *G. edulis*.

Name of Vitamins/Pigments	*G. corticata*	*G. edulis*
Vitamin B_1	0.38 ± 0.02	0.36 ± 0.02
Vitamin B_2	0.05 ± 0.01 [b]	1.54 ± 0.07 [a]
Vitamin B_3	1.54 ± 0.39	1.10 ± 0.29
Vitamin B_6	3.79 ± 0.30 [b]	4.77 ± 0.23 [a]
Vitamin B_9	1.00 ± 0.07 [a]	0.45 ± 0.06 [b]
Vitamin C	14.66 ± 0.23	13.41 ± 0.57
Vitamin A	2.67 ± 0.30 [a]	2.07 ± 0.06 [b]
Vitamin E	1.40 ± 0.10	1.49 ± 0.10
chlorophyll *a*	8.96 ± 0.39 [b]	17.14 ± 0.55 [a]
chlorophyll *b*	7.74 ± 0.33	8.44 ± 0.63
carotenoid	12.82 ± 0.50 [a]	2.99 ± 0.56 [b]

Values are mean ± standard deviation, $n = 3$ on dry weight basis. [a–b] values with different superscripts within the same line are significantly different.

The methanolic extracts of *G. corticata* and *G. edulis* at 1 mg/mL concentration indicate the presence of three major compounds, chlorophyll *a*, chlorophyll *b* and carotenoids, in *G. corticata* (Retention factor (Rf) value = 0.97, 0.92 and 0.95, respectively) and *G. edulis* (Rf = 0.96, 0.96 0.84, respectively). *G. corticata* and *G. edulis* contained 8.96 ± 0.39 µg/g and 17.14 ± 0.55 µg/g of chlorophyll *a* and 7.74 ± 0.33 µg/g and 8.44 ± 0.63 µg/g of chlorophyll *b*, respectively. With respect to the carotenoid content, it was higher for *G. corticata* (12.82 ± 0.50 µg/g) than for *G. edulis* (2.99 ± 0.56 µg/g).

Table 6 shows the swelling capacity (SWC), water-holding capacity (WHC) and oil-holding capacity (OHC) of *G. corticata* and *G. edulis*. In general, as temperature varied, the SWC and WHC of *G. corticata* and *G. edulis* powder varied, due to an increase in the solubility of the dietary fiber and the presence of protein in *G. corticata* and *G. edulis*. However, it also reaches significant differences for the case of SWC, whereas no statistical differences were obtained for WHC or OHC. The SWC of *G. edulis* were higher than *G. corticata* at both 25 °C and 37 °C (8.66 ± 0.53 mL/g vs. 7.90 ± 0.32 mL/g, and 7.70 ± 0.60 mL/g vs. 5.70 ± 0.65 mL/g, respectively).

Table 6. Swelling capacity (SWC), water holding capacity (WHC) and oil holding capacity (OHC) of *G. corticata* and *G. edulis*.

Seaweeds	SWC (mL/g)		WHC (g/g)		OHC (g/g)	
	25 °C	37 °C	25 °C	37 °C	25 °C	37 °C
G. corticata	7.90 ± 0.32 [bA]	5.70 ± 0.65 [bB]	4.03 ± 0.39	3.96 ± 0.58	2.06 ± 0.24	1.84 ± 0.40
G. edulis	8.66 ± 0.53 [aA]	7.70 ± 0.6 [aB]	4.09 ± 0.32	3.64 ± 0.40	1.87 ± 0.28	1.91 ± 0.18

Values are mean ± standard deviation, $n = 3$ on dry weight basis. [a–b] values with different superscripts within the same column are significantly different between seaweeds. [A–B] values with different superscripts within the same line shows significant differences between temperatures.

With respect to the WHC of *G. corticata* and *G. edulis*, values of 4.03 ± 0.39 and 4.09 ± 0.28 g/g, respectively, were obtained at 25 °C, reduced to 3.96 ± 0.58 g/g in *G. corticata* and 3.64 ± 0.18 g/g in *G. edulis* at 37 °C. In this study, both *G. corticata* and *G. edulis* exhibited similar OHC values (about 2 g/g) at both 25 °C and 37 °C.

3. Discussion

With respect to proximate content, the moisture of *G. corticata* and *G. edulis* was lower than most results obtained for *Gracilaria* sp. in general, such as the 12.15 g/100 g obtained for *G. acerosa* [6] the 19.2 g/100 g for *G. edulis* [8], and the 12.86 g/100 g for *G. edulis* [1], but higher than the 5.32 g/100 g obtained for *G. changii* [19]. In this work, *G. edulis* showed a higher ash content in a DW basis than *G. corticata*. Similarly, it was reported an ash content of 8.70 g/100 g in *G. edulis* [8], whereas other authors reported a higher ash content (40.30 g/100 g) in *G. changii* [19] than those found in the present work. A high ash content shows the presence of appreciable amounts of diverse minerals found in both seaweeds. A similar observation was for *G. changii* [19] in which were found an ash content of 6.40 g/100 g. Interestingly, total dietary fiber is known to have physiological properties for the prevention and treatment of cancer, obesity and diabetes [20,21]. Therefore, *G. corticata* and *G. edulis* may have the potential to be used as a source of dietary fiber in the nutraceutical industries.

Other authors found much lower crude protein contents in *Gracilaria* spp. than those found in the present work. Thus, it was reported a crude protein content of 6.68 g/100 g for *G. edulis* [8], 0.61 g/100 g for *G. acerosa* [6], 12.57 g/100 g in *G. changii* [19] or 19.70 g/100 g for *G. cervicornis* [22]. Moreover, the high protein content of *G. corticata* and *G. edulis* indicates that these seaweeds may be considered as potential marine plant sources of protein [22]. Proteins from seaweeds can have antibacterial, antioxidant, immunostimulating, antithrombotic and anti-inflammatory activities. Consequently, they can be used for prevention and treatment of hypertension, diabetes and hepatitis among other positive effects in the organism [20].

The total carbohydrate content of both *G. corticata* and *G. edulis* was markedly lower than that reported [8] for *G. edulis* (10.2 g/100 g) or the 29.44 g/100 g reported for *G. changii* [19]. However, other authors found lower carbohydrate content in *Gracilaria* species, such as *G. acerosa*, for which was reported a carbohydrate content of 1.05 g/100 g [6]. The wide variation in the carbohydrate content observed in red and brown seaweed species might be due to the influence of different factors like salinity, temperature and sunlight intensity [2]. Moreover, carbohydrate content is also influenced by biomass, which reveals the link between growth and carbohydrate content [23].

In general, seaweeds have a low fat content [23]; that makes seaweeds low-calorie foods and in the present work both seaweeds contained fat amounts of 7.07 g/100 g DW seaweed (*G. corticata*) and 4.71 g/100 g DW seaweed (*G. edulis*). These results are lower than those obtained by other authors [8], whose reported a crude lipid content for *G. edulis* of 8.30 g/100 g but significantly that the 0.3% reported for *G. changi* [19], or the 1.7–3.6% reported for *G. fisheri* and *G. tenuistipitata* [5]. Thus, *Gracilaria* content in fat can widely vary depending on the species and source.

Polysaccharides are polymers composed of at least 10 monosaccharides linked by glycosidic bonds [9]. Recently, seaweed polysaccharides have been given large attention by the scientific community due to their outstanding bioactivities and correspondingly low toxicity [9]. They have been shown to have other beneficial health effects, including their prebiotic effect and antioxidant or anti-inflammatory activity [20]. The polysaccharide content obtained in the present work was higher than the polysaccharide extracted from *Gracilaria* species in previous works, such as 29.08 g/100 g [24], 27.20 g/100 g [25], 21.40 g/100 g [26], and 32.80 g/100 g [27]. Contrariwise, it was also reported a higher polysaccharide content in *G. debilis* [28], in the range 52–67 g/100 g. A previous work [29] reported that the polysaccharide yield from *Gracilaria* species varies due to seasonal variations, physiochemical factors, environmental conditions and extraction methods. Additionally, the variations in the polysaccharide content of *Gracilaria* can vary depending on atmospheric temperature at the time of extraction [26]. Hence the present study significantly indicates that the crude polysaccharides present in *G. corticata* and *G. edulis* may exert varied biological activity [25].

With respect to the fatty acids composition, those of seaweeds often differ from those of terrestrial plants whereby seaweeds have a higher proportion of PUFAs than terrestrial vegetables. Red seaweeds are particularly rich in SFAs and PUFAs which have nutritional applications that lead to their extensive use in food, feed, cosmetic, biotechnological and pharmaceutical applications [30,31]. Variation in fatty acid content may also be due to the season of collection as well as other abiotic factors such as nutrition, salinity, light and temperature [8,20]. In the present work, total fatty acids were significantly lower than those obtained by other authors [8], who found 11.41 g/100 g in *G. edulis*. According to this work [8], the most abundant fatty acids in both seaweeds were palmitic, stearic and α-linoleic acid acids. The same fatty acids were also found abundant in *G. changii* [20]. However, our results were significantly lower than those obtained in *G. changii* for DHA content, in which DHA were found as the most abundant fatty acids, with a 48.36% of total fatty acids. The results of the present study revealed that both seaweeds are rich in SFAs and especially in PUFAs, which provide important health benefits. With respect to most commonly found n-3 PUFA, eicosapentaenoic acid (EPA) and DHA, it is common that their contents vary dramatically from *Gracilaria* spp. and even into the same species [32]. No EPA presence were found for the seaweeds tested in the present work. The presence of this n-3 fatty acid in *Gracilaria* spp. is inconstant, because it was found in *G. gracilis* [20], but it was not detected in *G. changii* [19] or *G. edulis* [8]. Fatty acids overall profile obtained in this work were significantly different than 57.5% SFAs, 18.3% MUFAs and 18.4% PUFAs reported for *Gracilaria* sp. [3] or the 7.5% SFAs, 38.3% MUFAs and 51.2% PUFAs 18.4% reported for *Gracilaria changii* [19].

The protein composition found in this work for *G. corticata* and *G. edulis* was lower than those found in a previous work [20], which reported an amino acid content of 91.90 mg/g in *G. changii*. The EAAs/total amino acid ratio was higher than those previously reported [6,19,26]. Aspartic acid content, that is important for the organoleptic point of view because it was reported that it is responsible for the special flavor and taste of seaweeds [33], was in similar contents in both seaweeds.

Seaweeds are one of the richest sources of minerals and trace elements, because the cell-wall polysaccharides and proteins of seaweed contain sulfate, anionic carboxyl and phosphate groups which act as binding sites for metal retention [34]. With respect to the mineral content, *G. corticata* showed a higher content of Mg (463.23 mg/kg) and Fe (1072.48 mg/kg) than *G. edulis*. Moreover, *G. edulis* was found to possess more of trace elements like Zn (42.73 mg/kg) and Cu (14.61 mg/kg) than *G. corticata*. Both seaweeds had a higher or similar content of minerals like Zn, Cu, Mg and Fe when compared with the content of *G. acerosa* [6], *G. edulis* [8], *G. fisheri* and *G. tenuistipidatata* [5] or *G. changii* [19], with the exception of Mg in *G. edulis* which were lower than those found for other previous works as *G. changii* [19]. The ability of seaweeds to accumulate metals will depend on a variety of factors such as location, exposure, salinity, temperature, pH, light, nitrogen content, season, plant age, metabolic processes or the affinity of the plant for each element among others [35]. In view of the present results, both *G. corticata* and *G. edulis* contain an adequate amount of minerals, which suggests that these seaweeds could act as important sources of mineral supplements which are essential for human nutrition.

For both *G. corticata* and *G. edulis,* the presence of water-soluble vitamins (vitamin B_1, vitamin B_2, vitamin B_3, vitamin B_6, vitamin B_9 and vitamin C) and fat-soluble vitamins (vitamin A and vitamin E) was found. *G. corticata* had a higher vitamin A (2.67 mg/g vs. 2.14 mg/g) and vitamin B_9 contents (1.00 mg/g vs. 0.45 mg/g) than *G. edulis*, whereas *G. edulis* showed a significantly higher content of vitamin B_2 (1.54 mg/g vs. 0.05 mg/g) and vitamin B_6 (4.77 mg/g vs. 3.79 mg/g) than *G. corticata*. With respect to previously published works, the vitamin content reported for *Gracilaria* species is widely different between the different authors [1,6,8]. Perhaps the more remarkable difference in vitamin content is that in the present work both *G. corticata* and *G. edulis* showed a significantly higher vitamin A content (2.67 and 2.07, respectively) than those previously reported for *G. acerosa* [6] or for *G. edulis* [1].

Another important difference was found for the case of vitamin C that showed a higher content than those previously reported [6,8] for *G. edulis* or *G. acerosa*, respectively. The variation in vitamin content may be due to some environmental factors such as salinity, atmospheric temperature, seasonality and methods of preservation and processing [6].

G. corticata and *G. edulis* contained 8.96 µg/g and 17.14 µg/g of chlorophyll *a* and 7.74 µg/g and 8.44 µg/g of chlorophyll *b*, respectively. In a previous work describing the composition of several seaweeds [36] it was reported that chlorophyll *a* and *b* in red seaweeds ranged from 68 to 162 μg/g and from 25 to 46 μg/g, respectively. Specifically, for *Gracilaria* spp., it was reported high chlorophyll *a* of 577.89 μg/g and low chlorophyll *b* of 1.11 μg/g in *G. changii* [19].

With respect to carotenoid content, a higher total carotenoid content was reported for *G. changii* than in the present study (74.22 μg/g) [19]. Carotenoids such as *β*-carotene, lutein, zeaxanthin and antheraxanthin have been identified in red seaweed, including *Gracilaria* species. Further, seaweed carotenoids, especially *β*-carotene, are preferred by the market of natural products, because they are a mixture of cis and trans isomers, which may possess anticancer activity [37].

The SWC, WHC and OHC properties of seaweeds are generally related to their content and type of polysaccharides as well as protein which links to the cell wall of polysaccharide [5]. Previous works described that variations in temperature can widely vary physicochemical properties of seaweeds, due to increase in the solubility of the dietary fiber and the presence of protein [5,8]. However, in our work, only were found significant variations in the case of SWC. Previous works [8] reported a SWC of 20 mL/g in *G. edulis*, higher than those found in the present study. Similarly, a SWC at 37 °C of 7.68 mL/g in *G. changii* was reported [19], whereas for *G. acerosa* [6] an SWC at 37 °C of 5 mL/g was reported.

With respect to the WHC of *G. corticata* and *G. edulis*, a similar observation was also made in a previous work [8], that reported a WHC for *G. edulis* of 3.08 g/g. Other authors found better WHC than in the present work for other *Gracilaria* spp., such as *G. fisheri*, for which a WHC of 5.53 g/g was reported [5], and *G. changii* [26], for which WHC values of 6.15 g/g at 24 °C and 9.93 g/g at 37 °C were reported. Both SWC and WHC of seaweeds might be attributed due to different protein content and increases in the number and nature of the water binding sites on the protein molecules [38].

OHC is another functional property of food ingredients used in formulated foods for consumption. Ingredients with high OHC values allow the stabilization of food emulsions and high-fat food products [19]. For other *Gracilaria* spp., it was reported [8] that *G. edulis* showed an OHC of 1.64 g/g, which is very similar to the OHC values of the present study. Moreover, for *G. changii* [19] OHCs of 3.11 g/g at 24 °C and 1.17 g/g at 37 °C were reported. The low oil absorption capacity of red seaweeds is generally related to the hydrophilic nature of the changed polysaccharides (agar, carrageenan, fucans and alginates) of soluble dietary fiber [39]. The results of the present study for physicochemical properties confirmed that *G. corticata* and *G. edulis* could be considered as a source of food ingredients including proteins, dietary and soluble fiber [39].

4. Materials and Methods

4.1. Sample Collection

Samples of the commercially important and commonly edible red seaweeds *G. edulis* and *G. corticata* were collected by hand from the Thondi Coast (Latitude: 9° 44′ N and Longitude: 79° 00′ E), Palk Bay, on the southeast coast of India. Freshly collected seaweeds were washed thoroughly in seawater and transported to the laboratory immediately. Epiphytes, sediment particles and other debris were removed by washing thoroughly using potable water and thoroughly washed with distilled water immediately after washing with potable water. Seaweeds were identified using a standard manual [40]. The voucher specimen was deposited in museum at Department of Oceanography and Coastal Area Studies, School of Marine Sciences, Alagappa University. Seaweeds were shade dried for 5 days at constant temperature of 25 °C ± 2 °C. Dried seaweed samples were powdered using a mechanical blender and stored at room temperature in an airtight container (Tarsons, Kolkatta, India) for further analysis within a maximum period of one week. Further the remaining powder sample stored at frozen condition (−20 °C) for future use.

4.2. Proximate Composition

The proximate composition of each *Gracilaria* species was determined in all cases following Association of Official Analytical Chemists (AOAC) methods [41]. AOAC methods were employed to determine in *G. corticata* and *G. edulis* the ash content by heating at 550 °C for 24 h in a muffle furnace (AOAC, 930.05), moisture content by heating at 105 °C for 24 h (AOAC, 934.01), total fat by Soxhlet extraction with petroleum ether (AOAC 991.36) and protein by the Kjeldahl method (N × 6.25) (AOAC 981.10) [41]. The total carbohydrate content of *G. corticata* and *G. edulis* was determined by the phenol–sulfuric acid method [38]. All measurements were performed in triplicate for each seaweed and expressed as g/100 g seaweeds in a dry weight matter.

4.3. Isolation of Polysaccharides

Polysaccharides were separated from *G. corticata* and *G. edulis* using 2 g samples of each seaweed according as previously described [42]. Powdered *G. corticata* and *G. edulis* were dissolved and homogenized with distilled water under constant stirring for 2 h at 100 °C. The residues obtained were then removed by centrifuging the sample at 6300 *g* for 10 min using a CPR 30 plus centrifuge (Remi Lab World, Mumbai, India). The obtained supernatant was then precipitated by addition of an ethanol-in-water solution (1:3 v/v), followed by subsequent washing with 30 mL of acetone. The precipitated polysaccharides were then collected and subsequently air-dried, re-dissolved in distilled water and washed with acetone. Afterwards, the collected polysaccharides were stored at −20 °C in a deep freezer (Blue Star, Mumbai, India) for further use. Isolation were performed in triplicate for each seaweed and results are expressed as g/100 g seaweeds in a dry weight basis.

4.4. Extraction of Crude Lipid and Determination of Fatty Acids Content

Portions of 500 mg each of powdered *G. corticata* and *G. edulis* were mixed with 5 mL of a chloroform:methanol solution (2:1 v/v), tightly covered with aluminum foil and kept at room temperature for 24 h. After this period, solutions were filtered through 11 μm Whatman No. 1 filter paper, and the filtered extract was placed in a pre-weighed and oven-dried beaker. The beaker was weighed with lipids, and the difference in weight was taken as total lipid content and expressed as a percentage [43]. Afterwards, an aliquot of the total lipids of each sample was used to determine the fatty acids content, based on a method published [6]. For this purpose, 0.45 g was introduced into a 10 mL volumetric flask, dissolved in hexane containing 50 mg of butylated hydroxytoluene per L and diluted to 10 mL with the same solvent. Afterwards, 2 mL of the solution was transferred into a quartz tube and evaporated by means of a nitrogen flow. Further, 1.5 mL of a 20 g/L solution of sodium hydroxide in methanol, covered with nitrogen, was added, capped tightly with a polytetrafluoroethylene-lined cap, mixed and heated in a water bath for 7 min. After the water bath, samples were cooled at room temperature, and 2 mL of boron trichloride-methanol solution was added; then, they were blanketed with nitrogen, capped tightly, mixed and heated in a water bath for 30 min. After this period, samples were cooled to 40–50 °C, and 1 mL of trimethylpentane was added; then, they were capped and shaken vigorously for at least 30 s. Immediately, 5 mL of saturated sodium chloride solution was added, then the samples were covered with nitrogen, capped and vortexed or shaken thoroughly for at least 15 s. The upper layer was allowed to become clear and then transferred to a separate tube. In the separate tube, the methanol layer was shaken once more with 1 mL of trimethylpentane and combined with trimethylpentane extracts. The organic solvent was then removed, and FAME were subjected to gas chromatography (GC), performed on a Perkin Elmer Clarus 580 gas chromatograph (Perkin Elmer, Gaithersburg, MD, USA) equipped with a flame ionization detector and an HP-5 capillary column (30 m × 0.25 mm). Initial temperature was maintained at 70 °C, then increased to 250 °C (10 °C/min); the injection temperature employed was 225 °C. Helium was used as carrier gas, with a flow rate of 1 μL/min. FAME peaks were identified by comparison of their retention times and quantified by comparison with individual calibration curves performed with a standard FAME mix (Supelco, Sigma-Aldrich, St Louis, MO, USA). Tricosanoic acid (C23; Sigma-Aldrich) was used as internal standard. Fatty acid composition was determined in triplicate for each seaweed and was expressed as g FAME/100 g total fat.

4.5. Protein Composition

Protein composition of *G. corticata* and *G. edulis* was determined based on a reversed-phase high performance liquid chromatography (HPLC) analysis method [6]. Two grams each of powdered seaweeds was mixed with phosphate buffer (pH 7.0) and centrifuged at 1200 g (Remi Lab World) for 20 min at 4 °C. The supernatant was collected, and protein content was precipitated by adding 10% v/v trichloroacetic acid (TCA). The protein pellet was resuspended in 1 N NaOH and hydrolyzed by heating the solution with 6 N HCl in a boiling water bath for 24 h. After incubation, the supernatant was collected by centrifuging the sample at 3500 g for 15 min. The supernatant was then filtered and neutralized by the addition of 1 N NaOH. The filtered supernatant was diluted to 1:100 (v/v) with deionized water. The sample was subjected to reversed-phase HPLC analysis (Lachrome Hitachi, Tokyo, Japan) with UV and fluorescence detectors. One µl of sample was injected into a Denali C18 5-mm column (4.6 mm × 150 mm) at 23 °C with detection at 254 nm. The mobile phase used was 20 mM sodium acetate/triethylamine (0.018% v/v) in phase A. The pH was adjusted to 7.2 using 1–2% acetic acid. In phase B, 20% of 100 mM sodium acetate (pH 7.2) with 1–2% acetic acid was used; 40% acetonitrile was used as the mobile phase. The protein composition was determined in triplicate for each seaweed and expressed as mg of amino acid/g of seaweed in a DW basis.

4.6. Determination of Mineral Content

Zn, Cu, Mg and Fe analysis of *G. corticata* and *G. edulis* was performed according to the European Standards, with minor modifications according to the method previously described [17]. One gram of homogenized seaweed sample was added to mixed reagent at a ratio of 5:2:1 (nitric acid:perchloric acid:sulfuric acid). Mineralization was performed on a hot plate at 50 °C for 30 min. After the end of digestion, 10 mL of 2 N HCL was added; digested solvents were filtered and made up to 25 mL with distilled water and stored at room temperature for further analysis [44]. Zn, Cu, Mg and Fe were determined by atomic absorption spectroscopy (AAS; Anton Paar-AAS, Graz, Austria). The metal standards were prepared and run to check the precision of the instrument throughout the analysis. Quality assurance and quality control protocols set by the US Environmental Protection Agency [45] for metal analysis were used. Quality assurance testing relied on the control of blanks and yield for the chemical procedure. Mineral content was determined in triplicate for each seaweed and expressed as mg/kg of DW seaweed.

4.7. Determination of Vitamin Content

The vitamin content of *G. corticata* and *G. edulis* was determined by the method previously described [20]; vitamin A or retinol (328 nm), vitamin B1 or thiamine monohydrate (420 nm), vitamin B2 or riboflavin (254 nm), vitamin B6 or pyridoxine HCl (254 nm), vitamin C or ascorbic acid, vitamin E or tocopheryl acetate (520 nm) and folic acid (550 nm) were determined by HPLC methods, and the results were compared with the respective standards retinyl acetate, thiamine monohydrate, riboflavin, pyridoxine HCl, ascorbic acid, tocopheryl, and folic acid (Sigma Aldrich). Vitamin content was determined in triplicate for each seaweed and was expressed as mg/g of DW seaweed.

4.8. Determination of Chlorophyll a, b and Carotenoids

Thin-layer chromatography (TLC) was used to screen chlorophyll *a* and *b* and carotenoid content in the *G. corticata* and *G. edulis* extracts [8]. The mobile phase contained methanol and chloroform (1:9). The sample (approximately 1 mg/mL) was spotted onto the TLC plates and air-dried. The spots were identified under long-wave and short-wave UV light, and also in an iodine chamber. The Rf value, which is the distance moved by the solute relative to the distance moved by the solvent, was calculated to find chlorophyll *a* and *b* and carotenoid content.

Afterwards, for chlorophyll *a* and *b* determination, 1 g each of *G. corticata* and *G. edulis* powder was extracted with 96% CH$_3$OH, and the supernatant was collected by centrifugation at 1000 g for 1 min.

After filtration, the supernatant was again filtered with Whatman No. 1 filter paper, centrifuged at 2300 g (Remi Lab World) for 10 min, and the absorbance of the collected supernatant was measured using a UV–Vis spectrophotometer [8]. The chlorophyll a and b contents were calculated using the following formulas in triplicate for each seaweed and expressed as µg/g of DW:

$$\text{Chlorophyll } a = 15.65 \,(A_{666}) - 7.340 \,(A_{653}) \tag{1}$$

$$\text{Chlorophyll } b = 27.05 \,(A_{653}) - 11.21 \,(A_{666}) \tag{2}$$

where A_{666} = absorbance at 666 nm; A_{653} = absorbance at 653 nm. Carotenoid content was determined according to [46]. One gram each of *G. corticata* and *G. edulis* powder was extracted with 5 mL of acetone and incubated in the dark for 45 min, and the supernatant was collected by centrifugation at 10,000 g for 5 min. The supernatant was then stored in a refrigerator, and the extraction repeated with acetone until it became colourless. The supernatant was pooled and made up to 10 mL with acetone, and the absorbance of the collected supernatant was measured at 450 nm using a UV–Vis spectrophotometer (Shimadzu, Kyoto, Japan). Carotenoid content was calculated using the following formula in triplicate for each seaweed and expressed as µg/g of DW:

$$\text{Carotenoid content} = A450/2500 \tag{3}$$

where 2500 is the extinction coefficient.

4.9. Physicochemical Properties

In order to determine the physicochemical properties of the *G. corticata* and *G. edulis* powder, the SWC, WHC and OHC were determined for each powder. The SWC of *G. corticata* and *G. edulis* was assessed based on a method previously described [38]. Briefly, 500 mg of *G. corticata* and *G. edulis* were taken and mixed with 20 mL of distilled water and stirred vigorously. The influence of temperature on SWC was determined by maintaining the tubes at 25 °C and 37 °C overnight. The SWC of *G. corticata* and *G. edulis* was calculated using the following formula and expressed as ml of swollen sample per g of DW:

$$\text{SWC} = \text{Initial volume of water (mL)} - \text{Volume of water after incubation (mL)} \tag{4}$$

The WHC of *G. corticata* and *G. edulis* was assessed by a modified method [38]. Briefly, 500 mg each of *G. corticata* and *G. edulis* was put into two sets of centrifuge tubes, and 20 mL of deionized water was added. The tubes were kept separately in an incubator shaker for 24 h at 25 and 37 °C. The supernatant was discarded after centrifuging the tubes at 12,000 g for 30 min (Remi Lab World). The wet weight of *G. corticata* and *G. edulis* was noted. The samples were dehydrated by keeping them in an oven at 160 °C for 2 h, and the dry weight of the sample was noted. The WHC was determined in triplicate for each seaweed and calculated using the following formula and expressed as the weight in g of water held by 1 g of DW sample:

$$\text{WHC} = \text{Wet weight of the sample (g)} - \text{Dry weight of the sample (g)} \tag{5}$$

The OHC of *G. corticata* and *G. edulis* was assessed according to previous methods [20,38]. About 3 g of dried *G. corticata* and *G. edulis* was taken in a tube and mixed with 10.5 g of corn oil. The tubes were placed in a shaker at room temperature for 30 min. The oil supernatant was collected by centrifugation at 3000 g for 30 min (Remi Lab World). The OHC was determined in triplicate for each seaweed and was determined using the following formula and expressed as the number of g of oil held by 1 g of DW sample:

$$\text{OHC} = \text{Initial volume of oil (g)} - \text{Volume of oil after incubation (g)} \tag{6}$$

4.10. Statistical Analysis

All results were expressed as mean ± SD. Paired sample *t*-test was used to compare composition values between *G. edulis* and *G. corticata*. One-way analysis of variance (ANOVA) and Duncan's test were used to compare the effects of temperature on the physicochemical properties. All determinations were performed using SPPS version 14 (SPSS Science, Chicago, IL, USA). A positive significant variation was defined at the significance level of $p < 0.05$.

5. Conclusions

The polysaccharide content of the investigated dried red seaweeds (*G. corticata* and *G. edulis*) were found to be rich sources of polysaccharides. Because of the potential as prebiotics, antioxidant and anti-inflammatory compounds, carbohydrates from seaweeds has are compounds in high demand by consumers today. The physicochemical properties and proximate composition revealed that both *G. corticata* and *G. edulis* have appreciable levels of ash, protein, carbohydrate, fatty acid, essential and non-essential amino acid and vitamin content. It is suggested that both the seaweeds tested have great potential as potential food supplements and may be used in the food industry as a source of ingredients with an appreciable amount of nutritional value. Since both red seaweeds were found to be a good source of essential nutrients, their commercial value can be enhanced by marketing them as value-added products. However, depending on their composition, *G. edulis* and *G. corticata* have important differences that make it more adequate for certain cases. *G. edulis* showed higher concentration of essential amino acids, chlorophyll, vitamin B_2 and Zn. Thus, it could be a good nutrient for low-protein diets or people whose need to reduce their oxidative status, because of its content in chlorophyll and Zn. Contrariwise, *G. corticata* showed higher PUFA content, carotenoids and minerals as Fe and Mg. Thus *G. corticata* is more adequate than *G. edulis* for people who need to reinforce their intake of such nutrients.

Author Contributions: Conceptualization, S.P. and A.A.; methodology, T.R. and A.A.; software, A.A. and A.M.-P.; validation, A.A.; formal analysis, T.R. and A.A.; investigation, A.A. and S.P.; resources, S.P.; data curation, J.M.M.; writing—original draft preparation, J.M.M.; writing—review and editing, A.M.-P.; visualization, S.P. and J.M.M.; supervision, S.P.; project administration, S.P.; funding acquisition, S.P.

Funding: This research received external funding by RUSA–2. 0 grant sanctioned vide Letter No. F. 24–51/2014–U, Policy (TNMulti–Gen), Department of Education, Government of India and by FEADER (The European Agricultural Fund for Rural Development (EAFRD)), grant number 2018/001B and ED431C 2018/05.

Acknowledgments: The authors thank the authorities of Alagappa University for providing necessary facilities.

Conflicts of Interest: The authors declare no conflict of interest.

References

1. Debbarama, J.; Rao, B.M.; Murthy, L.N.; Mathew, S.; Venkateshwarlu, G.; Ravishankar, C. Nutritional profiling of the edible seaweeds *Gracilaria edulis*, *Ulva lactuca* and *Sargassum* sp. *Indian J. Fish.* **2016**, *63*, 81–87. [CrossRef]
2. Torres, P.; Santos, J.P.; Chow, F.; dos Santos, D.Y. A comprehensive review of traditional uses, bioactivity potential, and chemical diversity of the genus *Gracilaria* (Gracilariales, Rhodophyta). *Algal Res.* **2019**, *37*, 288–306. [CrossRef]
3. Da Costa, E.; Melo, T.; Moreira, A.S.P.; Bernardo, C.; Helguero, L.; Ferreira, I.; Cruz, M.T.; Rego, A.M.; Domingues, P.; Calado, R.; et al. Valorization of lipids from *Gracilaria* sp. Through lipidomics and decoding of antiproliferative and anti-inflammatory activity. *Mar. Drugs* **2017**, *15*, 62. [CrossRef] [PubMed]
4. Meyer, B.J.; de Groot, R.H.M. Effects of omega-3 long chain polyunsaturated fatty acid supplementation on cardiovascular mortality: The importance of the dose of DHA. *Nutrients* **2017**, *9*, 1305. [CrossRef] [PubMed]
5. Benjama, O.; Masniyom, P. Biochemical composition and physicochemical properties of two red seaweeds (*Gracilaria fisheri* and *G. tenuistipitata*) from the Pattani Bay in Southern Thailand. *Songklanakarin J. Sci. Technol.* **2012**, *34*, 223–230.

6. Syad, A.N.; Shunmugiah, K.P.; Kasi, P.D. Seaweeds as nutritional supplements: Analysis of nutritional profile, physicochemical properties and proximate composition of *G. acerosa* and *S. wightii*. *Biomed. Prev. Nut.* **2013**, *3*, 139–144. [CrossRef]

7. Chan, P.T.; Matanjum, P.; Yasir, S.M.; Tan, T.S. Antioxidant and hypolipidaemic properties of red seaweed, *Gracilaria changii*. *J. Appl. Phycol.* **2014**, *26*, 302–310. [CrossRef]

8. Sakthivel, R.; Devi, K.P. Evaluation of physicochemical properties, proximate and nutritional composition of *Gracilaria edulis* collected from Palk Bay. *Food Chem.* **2015**, *174*, 68–74. [CrossRef]

9. Ju, T.; Deng, Y.; Xi, J. Optimization of circulating extraction of polysaccharides from *Gracilaria lemaneiformis* using pulsed electrical discharge. *ACS Sustain. Chem. Eng.* **2019**, *7*, 3593–3601. [CrossRef]

10. Moreda-Piñeiro, A.; Peña-Vázquez, E.; Bermejo-Barrera, P. Significance of the presence of trace and ultratrace elements in seaweeds. In *Handbook of Marine Macroalgae: Biotechnology and Applied Phycology*; Kim, S.K., Ed.; John Wiley & Sons: Chichester, UK, 2012; pp. 116–170.

11. Mišurcová, L. Chemical Composition of Seaweeds. In *Handbook of Marine Macroalgae: Biotechnology and Applied Phycology*; Kim, S.K., Ed.; John Wiley & Sons: Chichester, UK, 2012; pp. 173–192.

12. Jeon, T. Biological Effects of Proteins Extracted from Marine Algae. In *Handbook of Marine Macroalgae: Biotechnology and Applied Phycology*; Kim, S.K., Ed.; John Wiley & Sons: Chichester, UK, 2012; pp. 387–397.

13. Inanc, A.L. Chlorophyll: Structural properties, health benefits and its occurrence in virgin olive oils. *Akademik Gida.* **2011**, *9*, 26–32.

14. Hsu, C.Y.; Chao, P.Y.; Hu, S.P.; Yang, C.M. The antioxidant and free radical scavenging activities of chlorophylls and pheophytins. *Food Nutr. Sci.* **2013**, *4*, 1–8. [CrossRef]

15. Fleurence, J.; Morançais, M.; Dumay, J.; Decottignies, P.; Turpin, V.; Munier, M.; Garcia-Bueno, N.; Jaouen, P. What are the prospects for using seaweed in human nutrition and for marine animals raised through aquaculture? *Trends Food Sci. Technol.* **2012**, *27*, 57–61. [CrossRef]

16. Peinado, I.; Girón, J.; Koutsidis, G.; Ames, J. Chemical composition, antioxidant activity and sensory evaluation of five different species of brown edible seaweeds. *Food Res. Int.* **2014**, *66*, 36–44. [CrossRef]

17. Arulkumar, A.; Paramasivam, S.; Miranda, J.M. Combined effect of icing medium and red alga *Gracilaria verrucosa* on shelf life extension of Indian Mackerel (*Rastrelliger kanagurta*). *Food Bioprocess Tech.* **2018**, *11*, 1911–1922. [CrossRef]

18. Arulkumar, A.; Rosemary, T.; Paramasivam, S.; Babu Rajendran, R. Phytochemical composition, in vitro antioxidant, antibacterial potential and GC-MS analysis of red seaweeds (*Gracilaria corticata* and *Gracilaria edulis*) from Palk Bay, India. *Biocatal. Agric. Biotechnol.* **2018**, *15*, 63–71. [CrossRef]

19. Chan, P.T.; Matanjun, P. Chemical composition and physicochemical properties of tropical red seaweed, *Gracilaria changii*. *Food Chem.* **2017**, *221*, 302–310. [CrossRef] [PubMed]

20. Francavilla, M.; Franchi, M.; Monteone, M.; Caroppo, C. The red seaweed *Gracilaria gracilis* as a multi products source. *Mar. Drugs* **2013**, *11*, 3754–3776. [CrossRef]

21. Kendall, C.W.C.; Esfahani, A.; Jenkins, D.J.A. The link between dietary fibre and human health. *Food Hydrocoll.* **2010**, *24*, 42–48. [CrossRef]

22. Marinho-Soriano, E.; Fonseca, P.; Carneiro, M.; Moreira, W. Seasonal variation in the chemical composition of two tropical seaweeds. *Bioresour. Technol.* **2006**, *97*, 2402–2406. [CrossRef]

23. Mabeau, S.; Fleurence, J. Seaweed in food products: Biochemical and nutritional aspects. *Trends Food Sci Technol.* **1993**, *4*, 103–107. [CrossRef]

24. Seedevi, P.; Moovendhan, M.; Viramani, S.; Shanmugam, A. Bioactive potential and structural characterization of sulfated polysaccharide from seaweed (*Gracilaria corticata*). *Carbohydr. Polym.* **2017**, *155*, 516–524. [CrossRef]

25. Souza, B.W.; Cerqueira, M.A.; Bourbon, A.I.; Pinheiro, A.C.; Martins, J.T.; Teixeira, J.A.; Coimbra, M.A.; Vicente, A.A. Chemical characterization and antioxidant activity of sulfated polysaccharide from the red seaweed *Gracilaria birdiae*. *Food Hydrocoll.* **2012**, *27*, 287–292. [CrossRef]

26. Melo, M.; Feitosa, J.; Freitas, A.; De Paula, R. Isolation and characterization of soluble sulfated polysaccharide from the red seaweed *Gracilaria cornea*. *Carbohydr. Polym.* **2002**, *49*, 491–498. [CrossRef]

27. Barros, F.C.; da Silva, D.C.; Sombra, V.G.; Maciel, J.S.; Feitosa, J.P.; Freitas, A.L.; de Paula, R.C. Structural characterization of polysaccharide obtained from red seaweed *Gracilaria caudata* (J Agardh). *Carbohydr. Polym.* **2013**, *92*, 598–603. [CrossRef] [PubMed]

28. Sudharsan, S.; Subhapradha, N.; Seedevi, P.; Shanmugam, V.; Madeswaran, P.; Shanmugam, A.; Srinivasan, A. Antioxidant and anticoagulant activity of sulfated polysaccharide from *Gracilaria debilis* (Forsskal). *Int. J. Biol. Macromol.* **2015**, *81*, 1031–1038. [CrossRef] [PubMed]

29. Armisen, R. World-wide use and importance of *Gracilaria*. *J. Appl. Phycol.* **1995**, *7*, 231. [CrossRef]

30. Bocanegra, A.; Bastida, S.; Benedí, J.; Ródenas, S.; Sánchez-Muniz, F.J. Characteristics and nutritional and cardiovascular-health properties of seaweeds. *J. Med. Food* **2009**, *12*, 236–258. [CrossRef]

31. Kumari, P.; Kumar, M.; Gupta, V.; Reddy, C.R.K.; Jha, B. Tropical marine macroalgae as potential sources of nutritionally important PUFAs. *Food Chem.* **2010**, *120*, 749–757. [CrossRef]

32. Khotimchenko, S.V.; Vaskovsky, V.E.; Przhemenetskaya, V.F. Distribution of eicosapentaenoic and arachidonic acids in different species of *Gracilaria*. *Phytochemistry* **1991**, *30*, 207–209. [CrossRef]

33. Matanjun, P.; Mohamed, S.; Mustapha, N.M.; Muhammad, K. Nutrient content of tropical edible seaweeds, *Eucheuma cottonii*, *Caulerpa lentillifera* and *Sargassum polycystum*. *J. Appl. Phycol.* **2009**, *21*, 75–80. [CrossRef]

34. Ródenas de la Rocha, S.; Sánchez-Muniz, F.J.; Gómez-Juaristi, M.; Marín, M.T.L. Trace elements determination in edible seaweeds by an optimized and validated ICP-MS method. *J. Food Compos. Anal.* **2009**, *22*, 330–336. [CrossRef]

35. Sánchez-Rodríguez, I.; Huerta-Diaz, M.A.; Choumiline, E.; Holguín-Quiones, O.; Zertuche-González, J.A. Elemental concentrations in different species of seaweeds from Loreto Bay, Baja California Sur, Mexico: Implications for the geochemical control of metals in algal tissue. *Environ. Pollut.* **2001**, *114*, 145–160. [CrossRef]

36. Hong, D.D.; Hien, H.M.; Son, P.N. Seaweeds from Vietnam used for functional food, medicine and biofertilizer. *J. Appl. Phycol.* **2007**, *19*, 817–826. [CrossRef]

37. Christaki, E.; Bonos, E.; Giannenasa, I.; Florou-Paneria, P. Functional properties of carotenoids originating from algae. *J. Sci. Food Agric.* **2013**, *93*, 5–11. [CrossRef]

38. Wong, K.H.; Cheung, P.C.K. Nutritional evaluation of some subtropical red and green seaweeds Part II. In vitro protein digestibility and amino acid profiles of protein concentrates. *Food Chem.* **2001**, *72*, 11–17. [CrossRef]

39. Rupérez, P.; Saura-Calixto, F. Dietary fibre and physicochemical properties of edible Spanish seaweeds. *Eur. Food Res. Technol.* **2001**, *212*, 349–354. [CrossRef]

40. Jha, B.; Reddy, C.R.K.; Thakur, M.C.; Rao, M.U. The diversity and distribution of seaweeds of Gujarat Coast. In *Developments in Applied Phycology*; Springer: New York, NY, USA, 2009; pp. 1–213.

41. Association of Official Analytical Chemists (AOAC). *Official Methods of Analysis*, 17th ed.; AOAC: Gaithersburg, MD, USA, 2002.

42. Dubois, M.; Gilles, K.A.; Hamilton, J.K.; Rebers, P.A.; Smith, F. Colorimetric method for determination of sugars and related substances. *Anal. Chem.* **1956**, *28*, 350–356. [CrossRef]

43. Folch, J.; Lees, M.; Sloane Stanley, G.H. A simple method for the isolation and purification of total lipids from animal tissues. *J. Biol. Chem.* **1957**, *226*, 497–509.

44. *Manual of Methods in Aquatic Environment Research (pat 9). Analyses of Metals and Organochlorines in Fish*; FAO Fisheries Technical Paper; Food and Agriculture Organization of the United Nations: Rome, Italy, 1984; p. 212.

45. *Fish and Fishery Products Hazards and Controls Guidance, Center for Food Safety and Applied Nutrition's*, 3rd ed.; Food and Drug Administration (FDA): Gainesville, FL, USA, 2001.

46. Jansen, A. Chlorophylls and Carotenoids. In *Handbook of Phycological Methods*; Cambridge University Press: Cambridge, UK, 1978; pp. 60–70.

Sample Availability: Samples of the *Gracilaria edulis* and *Gracilaria corticata* are available from the authors.

Article

Concentration of EPA and DHA from Refined Salmon Oil by Optimizing the Urea–Fatty Acid Adduction Reaction Conditions Using Response Surface Methodology

Gretel Dovale-Rosabal [1], Alicia Rodríguez [1,*], Elyzabeth Contreras [1], Jaime Ortiz-Viedma [1,*], Marlys Muñoz [1], Marcos Trigo [2], Santiago P. Aubourg [2,*] and Alejandra Espinosa [3]

[1] Department of Food Science and Chemical Technology, Faculty of Chemical and Pharmaceutical Sciences, Santos Dumont 964, University of Chile, Santiago 8380000, Chile; gretel.dovale@ug.uchile.cl (G.D.-R); ely_tcn@hotmail.com (E.C.); marlys_dayane@hotmail.com (M.M.)

[2] Department of Food Technology, Marine Research Institute (CSIC), Eduardo Cabello, 6, 36208 Vigo, Spain; mtrigo@iim.csic.es

[3] Department of Medical Technology, School of Medicine, University of Chile, Santiago 8380000, Chile; bespinosa@med.uchile.cl

* Correspondence: arodrigm@uchile.cl (A.R.); jaortiz@uchile.cl (J.O.-V.); saubourg@iim.csic.es (S.P.A.)

Academic Editor: Jose M. Miranda
Received: 19 March 2019; Accepted: 20 April 2019; Published: 26 April 2019

Abstract: This research focused on obtaining eicosapentaenoic acid (EPA, 20:5 *n*-3) and docosahexaenoic acid (DHA, 22:6 *n*-3) (EPA+DHA) concentrates from refined commercial salmon oil (RCSO). Independent variables of the complexation process were optimized by means of the application of response surface methodology (RSM) in order to obtain the maximum content of such fatty acids (FAs). As a result of employing the optimized conditions for all the variables (6.0, urea:FA content ratio; −18.0 °C, crystallization temperature; 14.80 h, crystallization time; 500 rpm, stirring speed), high contents of EPA and DHA could be obtained from RCSO, achieving increases of 4.1 and 7.9 times in the concentrate, with values of 31.20 and 49.31 g/100 g total FA, respectively. Furthermore, a 5.8-time increase was observed for the EPA + DHA content, which increased from 13.78 to 80.51 g/100 g total FA. It is concluded that RCSO can be transformed into a profitable source of EPA and DHA (EPA+DHA), thus leading to a product with higher commercial value.

Keywords: refined commercial salmon oil; *n*-3 long-chain polyunsaturated fatty acids (*n*-3 LCPUFAs) concentration; EPA; DHA; EPA+DHA; total FA yield; process variable maximization; response surface methodology (RSM); multiple response optimization; desirability function

1. Introduction

In recent years, it has been recognized that the consumption of eicosapentaenoic acid (EPA) is associated with a low prevalence of coronary, circulatory, and inflammatory diseases [1–4]. Furthermore, docosahexaenoic acid (DHA) has been associated with fetal development, the prevention of neurodegenerative diseases, and the correct functioning of the nervous system and visual organs in the fetus [5–10]. According to the Food and Agriculture Organization/World Health Organization (FAO/WHO) [1], the recommended intake of EPA+DHA is at least 250 mg/day for adult males and non-pregnant/non-lactating adult females. Interestingly, the optimal brain development of children would need a 150 mg/day diet of such fatty acids. EPA+DHA concentrates may be produced by various methods, such as supercritical fluid chromatography, supercritical fluid fractionation, molecular distillation, silver complexation, enzymatic methods, and urea complexation [11,12]. Among them, complexation with the urea can be considered as the most efficient method, since polyunsaturated fatty acids (PUFAs) may be separated from saturated and monounsaturated ones by means of an economic process at low temperature [13–18]. The present research was focused on the employment of refined commercial salmon oil (RCSO) as a profitable source of EPA+DHA concentrates, which in time could lead to a product of higher commercial value. For it, independent variables of the urea adduction reaction conditions (urea:FA content ratio, crystallization time and temperature, and crystallization stirring speed) were optimized by response surface methodology (RSM) in order to achieve the maximum content of EPA, DHA and EPA+DHA. Additionally, the quality of the starting salmon oil was determined and evaluated.

2. Results

2.1. Characterization of the Initial Refined Commercial Salmon Oil

Results for the composition of RCSO reported that the most abundant fatty acids were 18:1 9c (29.61%), 18:2 9c, 12c (16.69%), and 16:0 (13.74 %) followed by EPA (7.53%) and DHA (6.25%) (g/100 g total FA). The total value of saturated fatty acids (SFAs) was 21.28%, which was composed mainly of palmitic, stearic, and myristic acids (Table 1). Special interest is the confirmation of the absence of phytanic acid within the lipid composition of RCSO. This fatty acid has been linked to neurological disorders in some people, but it is also associated with the prevention of metabolic syndrome or type 2 diabetes [19].

Table 1. Composition of fatty acids in (RCSO) and the optimized concentrate from RCSO (g/100 g total FA) *.

FA or FA Groups	RCSO	RCSO Optimum
12:0	0.07	Nd
14:0	3.19	0.12
15:0	0.20	0.09
16:0	13.74	Nd
16:1 9t	0.15	Nd
16:1 7c	Nd	Nd
16:1 9c	4.66	0.51
16:1 11c	Nd	Nd
16:1 13c	Nd	Nd
17:0	0.13	0.10
17:1 10c	0.56	1.28
18:0	3.69	0.28
18:1 9c	29.61	0.59
18:1 11c	3.69	0.04
18:2 9t, 12t	Nd	Nd
18:2 9c, 12c	16.69	7.48
18:2 9c, 15c	Nd	Nd
18:3 6c, 9c, 12c	0.22	1.09
20:0	0.26	Nd
18:3 9c, 12c, 15c	3.25	2.60
20:1 5c	Nd	Nd
20:1 8c	Nd	Nd
20:1 11c	1.60	Nd
18:4 6c, 9c, 12c, 15c	Nd	Nd
20:2 11c, 14c	0.79	0.05
20:3 8c, 11c, 14c	0.30	1.16
20:3 11c, 14c, 17c	0.12	0.03
20:4 8c, 11c, 14c, 17c	0.40	1.37
22:1 13c	0.21	Nd
20:5 5c, 8c, 11c, 14c, 17c	7.53	31.20
24:1 15c	Nd	Nd

Table 1. *Cont.*

22:5 7c, 10c, 13c, 16c, 19c	2.69	2.70
22:6 4c, 7c, 10c, 13c, 16c, 19c	6.25	49.31
Total SFAs	21.28	0.59
Total MUFAs	40.48	2.42
Total PUFAs	38.24	96.99
Total *n*-3PUFAs	20.45	87.21
Total *n*-3LCPUFAs	18.08	85.82
EPA+DHA	13.78	80.51

* Abbreviations employed: DHA (docosahexaenoic acid), EPA (eicosapentaenoic acid), FA (fatty acid), SFAs (saturated fatty acids), MUFAs (monounsaturated fatty acids), RCSO (refined commercial salmon oil), n-3LCPUFAs (n-3 long chain polyunsaturated fatty acids), Nd (not detected).

Values obtained for the oxidative stability of RCSO were: peroxide value (PV) = 5.23 ± 0.05 meq. active oxygen kg^{-1} oil; p-anisidine value (pAV) = 6.84 ± 0.46; and total oxidation value (TOTOX) = 17.30 ± 0.40 and free fatty acids (FFA) = 0.30 ± 0.01 g oleic acid/100 g oil. Previous studies performed on refined salmon oil samples [18] indicated average values of PV= 3.54 ± 0.16 meq active oxygen kg^{-1} oil; p-anisidine value (pAV) = 5.14 ± 1.02 and FFA= 0.23 ± 0.00 g oleic acid/100 g oil. and at and 12.22 ± 1.34, respectively.

2.2. Effect of Process Variables on Total FA Yield, EPA Contents, and DHA Contents of RCSO Concentrate

2.2.1. Refined Commercial Salmon Oil Concentrate

According to the experimental design, showed in Table 2, 28 assays were performed to obtain different refined commercial salmon oil concentrates. This table reports the experimental values obtained for the different response variables: R_1 (total FA yield; g FA in the non-urea complexing fraction/100 g initial saponified oil FA), R_2 (EPA content; g/100 g total FA), and R_3 (DHA content; g/100 g total FA) of RCSO concentrate. Table 2 also includes the predicted values for the described corresponding variables, whereas the experimental values were replaced by the application of the model (R1′, R2′, and R3′ values, respectively). As a result, all the independent variables (A: urea/FA content ratio, *w/w*; B: crystallization temperature, °C; C: crystallization time, h; D: stirring speed, rpm) significantly affected ($p < 0.05$) the response variables during the urea complexation process. On the other hand, the enrichment of EPA and DHA in concentrates varied inversely according to total FA yield, obtaining correlation coefficient values (r) of −0.7756 and −0.7185, respectively.

Table 2. Values obtained for the experimental and predicted response variables of RCSO concentrate by central composite rotatable design 2^4 + star based on the response surface methodology [1].

Run	Process Variables *				Response Variables **					
					Experimental Values			Predicted Values		
	A	B	C	D	R_1	R_2	R_3	R_1'	R_2'	R_3'
1	1.5	−15	14.3	200	40.44	10.47	11.04	33.77	13.77	14.00
2	4.5	−15	14.3	200	12.64	28.41	44.38	12.23	28.09	45.89
3	1.5	15	14.3	200	48.71	9.22	9.52	40.35	9.52	10.23
4	4.5	15	14.3	200	17.45	25.09	32.69	18.82	23.85	26.94
5	1.5	−15	36.8	200	36.25	10.46	11.13	29.02	13.22	15.24
6	4.5	−15	36.8	200	10.44	24.42	55.91	21.55	27.55	47.14
7	1.5	15	36.8	200	4.22	7.96	8.18	20.22	8.98	11.47
8	4.5	15	36.8	200	18.12	24.87	30.47	12.76	23.30	28.18
9	1.5	−15	14.3	600	41.74	10.23	10.89	44.89	14.52	14.71
10	4.5	−15	14.3	600	12.77	30.20	46.43	12.14	28.84	46.61
11	1.5	15	14.3	600	78.45	9.02	9.41	68.43	10.27	10.94
12	4.5	15	14.3	600	16.11	25.16	34.34	21.40	24.60	27.65

Table 2. *Cont.*

13	1.5	−15	36.8	600	45.49	9.91	10.06	40.14	13.97	15.96
14	4.5	−15	36.8	600	9.86	27.56	48.55	7.17	28.30	47.85
15	1.5	15	36.8	600	58.94	7.84	7.91	48.30	9.73	12.18
16	4.5	15	36.8	600	12.64	23.05	28.93	15.33	24.05	28.89
17	0	0	25.5	400	57.38	6.33	6.39	67.84	-0.77	0.94
18	6	0	25.5	400	15.86	25.45	30.43	13.34	27.88	49.55
19	3	−30	25.5	400	12.21	29.26	44.37	19.61	24.94	36.61
20	3	30	25.5	400	33.82	13.52	14.65	34.36	16.45	13.88
21	3	0	3.05	400	12.71	20.41	22.67	25.65	21.24	24.00
22	3	0	48.0	400	16.31	23.01	28.93	14.83	20.15	26.49
23	3	0	25.5	0	19.11	22.62	25.72	13.39	19.95	24.53
24	3	0	25.5	800	16.47	26.07	33.39	27.09	21.44	25.95
25	3	0	25.5	400	20.77	22.43	25.82	20.24	20.69	25.24
26	3	0	25.5	400	20.74	21.40	25.20	20.24	20.69	25.24
27	3	0	25.5	400	17.20	24.00	29.15	20.24	20.69	25.24
28	3	0	25.5	400	22.74	18.23	20.23	20.24	20.69	25.24

Equations	R^2 Adjusted
Total FA yield $= 57.60 - 19.47A + 0.26B - 0.87C + 0.08D + 2.26AA + 0.01BB + 0.21AC - 0.02AD - 0.02BC + 0.001BD$	0.72(1)
EPA $= -0.90 + 9.53A - 0.14B - 0.79AA$	0.84(2)
DHA $= -1.19 + 8.10A + 0.17B - 0.19AB$	0.81(3)
EPA+DHA $= 3.52 + 12.88A - 0.52B$	0.80(4)

[1] Central composite design: 2^4 + star, which studies the effects of 4 factors in 28 runs based on the RSM. * Independent variables: A (urea/FA content ratio, w/w), B (crystallization temperature, °C), C (crystallization time, h), and D (stirring speed, rpm).** Response variables: R_1 (total FA yield, g FA in the non-urea complexing fraction/100 g initial saponified oil FA), R_2 (EPA content, g/100 g total FA), and R_3 (DHA content, g/100 g total FA). Predicted response variables: $R_1{}'$, $R_2{}'$, and $R_3{}'$.

2.2.2. Effect of Process Variables on EPA, DHA, and EPA+DHA Content and Total FA Yield: Pareto Charts and RSM Analysis

Pareto charts (Figure 1) were obtained for the different dependent variables as a function of the concentrate processing variables from RCSO; furthermore, the linear, quadratic, and interaction terms in the second-order polynomial were used to generate a three-dimensional response surface graph. Panel (A) indicates that the total FA yield (R1) of concentrates was dependent ($p < 0.05$) on the linear terms of the urea:FA content ratio (A), crystallization temperature (B), crystallization time (C), stirring speed (D), the quadratic terms AA and BB, and the interactions terms AD, BD, BC, and AC. Figure 1 (Panel B) shows the response surface of the urea complexation process for the total FA yield. The total FA yield decreased when the urea:FA content ratio increased and the crystallization temperature decreased. A similar result was found for the total FA yield effect in concentrate obtained from a by-product of rainbow trout processing where the total FA yield presented a minimum value in the response surface analysis when considering high urea:FA content ratios, at low crystallization temperature levels and stirring speeds, and at intermediate levels of crystallization time [17]. In the case of the EPA content (Figure 1, Panel C), the Pareto charts showed that certain linear terms—the urea:FA content ratio (A), crystallization temperature (B), and the quadratic urea:FA content ratio (AA)—provided a significant effect ($p < 0.05$). However, linear terms such as the crystallization time (C) and stirring speed (D) did not produce significant changes ($p > 0.05$). Figure 1 (Panel D) exhibits the response surface of the urea complexation process for EPA content. It was found that the EPA content increased with the urea:FA content ratio while it decreased as the crystallization temperature increased. For the DHA content (Figure 1, Panel E), the Pareto charts reported that the linear terms of the urea:FA content ratio (A), crystallization temperature (B), and the interaction between the urea:FA content ratio and the crystallization temperature (AB) revealed a significant effect ($p < 0.05$). The DHA content increased when the urea:FA content ratio increased, while it decreased as the crystallization temperature increased (Figure 1, Panel F). Finally, concerning the EPA+DHA content, the urea:FA content ratio showed a positive effect, whereas the crystallization temperature had a negative effect ($p < 0.05$) (Figure 1, Panel G). These results agree with those obtained by authors concerning the employment of the Asian catfish (*Pangasius bocourti*) and by-product of rainbow trout oil, which reported an inverse relationship between the urea:FA content ratio and the crystallization temperature on the urea concentrate process [16,17]. When the urea:FA content ratio increased and the crystallization temperature decreased as result, high values were obtained for DHA in the non-urea complexing fraction, as well as a great retention of saturated and monounsaturated FA in the urea crystal adducts. In this case, similar results for the EPA content were found. In all the cases, the EPA, DHA, and EPA+DHA contents did not significantly varied depending on the stirring speed ($p > 0.05$).

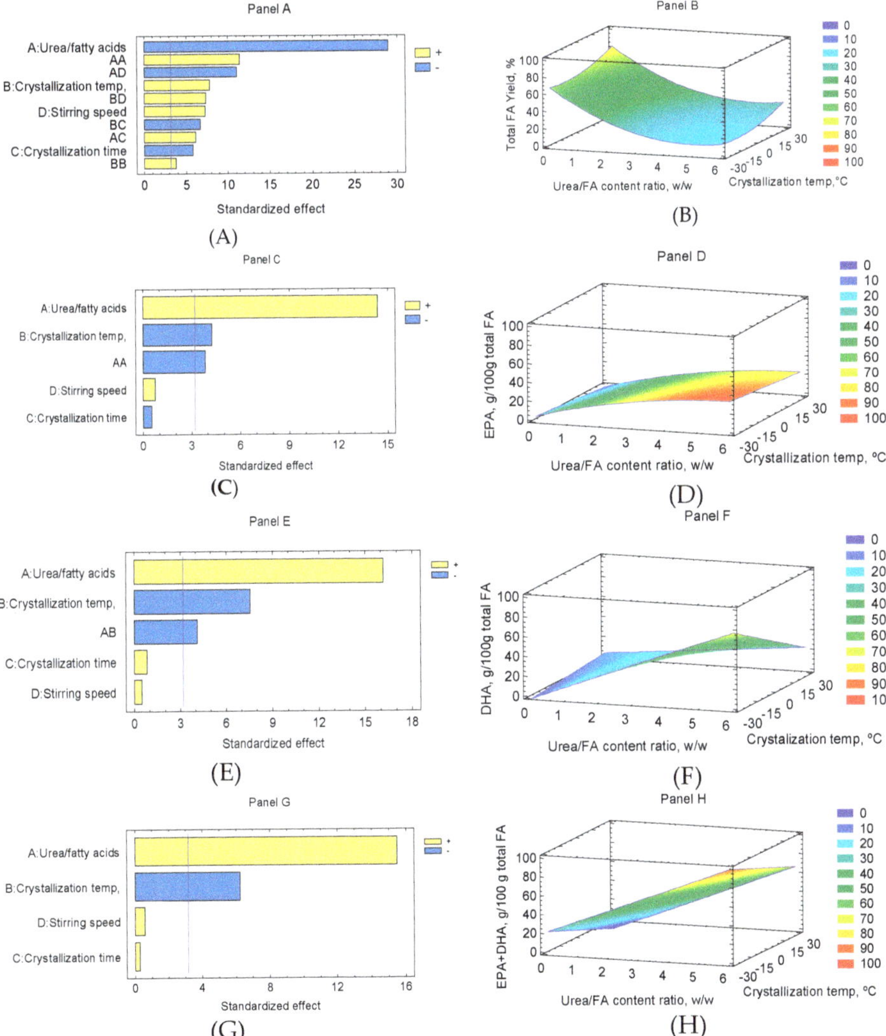

Figure 1. Pareto charts and response surfaces for the effects of different process variables: Panels in total FA yield (%, panels **A,B**), EPA content (g/100 g total FA, panels **C,D**), DHA content (g/100 g total FA, panels **E,F**), EPA+DHA content (g/100 g total FA, panels **G,H**). A: urea/FA contents ratio, *w/w*; B: crystallization temperature, °C; C: crystallization time, h; and D: stirring speed, rpm).

2.2.3. Models Obtained for the Concentration of EPA, DHA, and EPA+DHA

Equations obtained for the experimental process variables of the response surface model calculated by multiple regression are shown in Table 2. According to the equations obtained, all the response variables were found to be dependent on the same process variables expressed in the Pareto analysis (Figure 1). The four RCSO concentrated models had an adjusted R^2 by degrees of freedom of 72.0% for total FA yield (Equation (1)), 84.0% for EPA (Equation (2)), 81.0% for DHA (Equation (3)), and 80.0% for EPA+DHA (Equation (4)). Such values reported that the models adequately represented the variability of the results. Since the *p*-values obtained for lack-of-fit in the ANOVA study (0.014, 0.35, 0.19, 0.28; Equation (1) to Equation (4), respectively) was greater or equal to 0.05, except for Equation (1), the model appears to be adequate for the observed data at the 95.0% confidence level.

2.2.4. Independent Variables and Multiple Response Optimization

Table 3 (part a) shows the optimization of the independent variables for the response variables (EPA, DHA, and EPA+DHA) of the urea complexation process. The optimum values of dependent variables for EPA, DHA, and EPA+DHA content were 33.01, 76.81, and 98.85 (g/100 g total FA), respectively. In all the cases, a tendency toward the same values for the urea:FA content ratio and crystallization temperature was observed, so that the optimum EPA, DHA, and EPA+DHA contents were obtained.

Table 3. Process variables * optimization and multiple response optimization of the response variables.

Part a) Optimization of the Process Variables						
Dependent Variables	**Process Variables**				**Stationary Point**	**Optimum Value ******
	A	B	C	D		
EPA	5.99	−29.79	3.05	599.00	Maximum	33.01
DHA	6.00	−29.98	48.05	108.30	Maximum	76.81
EPA+DHA	6.00	−29.95	47.79	271.36	Maximum	98.85

Part b) Multiple Response Optimization of the Response Variables						
Dependent Variables	**Process Variables**				**Stationary Point**	**Predicted Value ******
	A	B	C	D		
EPA						30.71-
DHA	5.84	−17.69	14.83	453.36	Maximum	62.94
EPA+DHA						90.07
Maximum desirability						1.0

Part c) Experimental Validation of the Multiple Response Optimization of the Dependent Variables						
Dependent Variables	**Process Variables**				**Stationary Point**	**Experimental Value ******
	A	B	C	D		
EPA						31.20
DHA	6.00	−18	14.80	500	Maximum	49.31
EPA+DHA						80.51

* Process variables (A, B, C, and D) as expressed in Table 2. ** Values expressed as g/100 g total FA.

Table 3b shows the levels of factors that maximized the EPA, DHA, and EPA+DHA contents (g/100 g total FA) by means of multiple response optimization and the stationary point that predicted a maximum of 30.71, 62.94, and 90.07 (g/100 g total FA) in EPA, DHA, and EPA+DHA, respectively. Maximum desirability score of 1 (range, 0–1) was attained. A maximum predicted value could be obtained, provided that the following process conditions were applied: 5.84 (urea: FA content ratio), −17.69 °C (crystallization temperature), 14.83 h (crystallization time), and 453.00 rpm (stirring speed).

Figure 2 (panels A, B) shows the contours and estimated response surface of the urea:FA content ratio and crystallization temperature of the combination of factors levels to maximize the desirability function for RCSO concentrate. It is observed that the highest desirability values were reached by taking into account the high values of the urea:FA content ratio and the low crystallization temperature values. These results were similar to those previously reported for rainbow trout belly oil [17], where a maximum desirability score of 0.91 was obtained in the multiple response optimization of EPA, DHA and EPA+DHA contents. Furthermore, the predictive values of 32.50, 37.00, and 67.70 (g/100 g total FA) in the by-product concentrate of rainbow trout oil, respectively, were obtained. As a result, the following process conditions were applied in such study: 4.21 (urea:FA contents ratio), −15.00 °C (crystallization temperature), 24.0 h (crystallization time), and 1000 rpm (stirring speed).

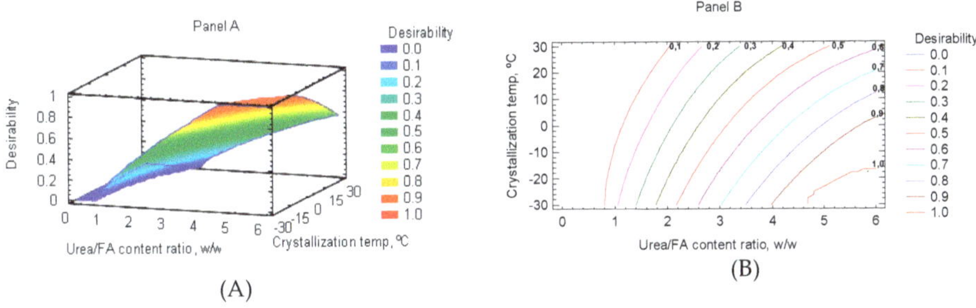

Figure 2. Combination of factors to maximize the desirability function for RCSO concentrate: response surface (**A**) and contour surface (**B**).

2.2.5. Validation of the Optimized Process and Characterization of the EPA+DHA Content Obtained

Table 3c shows the validation of multiple response optimization after experimentally performing the process conditions. For EPA content, comparison of the predicted value and the value obtained experimentally showed that both values were similar (i.e., 30.71 and 31.20 g/100 g total FA, respectively); however, the experimental value of DHA content was substantially different from the predictive value (49.31 and 62.94 g/100 g total FA, respectively). Experimental values performed on rainbow trout belly oil [17] were shown to be similar to the predicted values obtained in the current study for EPA and DHA (36.10 and 47.70 g/100 g of total FA, respectively). In the experimental validation, an 80.51 g/100 g total FA value was obtained for the EPA+DHA content, which agrees with the value revealed by other authors who obtained a stationary point of 89.38% for the EPA+DHA content [14].

2.3. Composition of FA in the RCSO Optimized Concentrate after Validation

Table 1 shows the FA composition of the RCSO compared to the composition of FA from the validated optimum concentrate. Optimization process validation was performed experimentally by combining the factors levels in which the optimal EPA+DHA content was attained (Table 3b,c). In these experiments, the effect of the urea complexation process on the composition of FA and FA groups in the optimized concentrate (g/100 g FA) after validation can be observed. When compared to the initial RCSO, there was a marked increase in the concentration of total PUFAs (96.99%) and a substantial decrease in the concentration of SFAs (0.59%) and MUFAs (2.42%). Furthermore, the predominant FAs in the optimum concentrate oil were EPA, DHA, and EPA+DHA (31.20, 49.31, and 80.51 g/100 g total FA, respectively). The urea complexation under the optimization process conditions has revealed a high efficiency, since a marked increase of total PUFAs from 38.24 to 96.99 g/100 g total FA could be observed; this increase was 1.21 times higher than that reported by Pando et al. [18]. Additionally, the total EPA+DHA final content was 2.5 times higher than that reported for refined salmon oil concentrate without the optimization process [18].

3. Discussion

In the present research, the distribution of SFAs in RSCO showed that this FA group was mainly composed of palmitic, stearic, and myristic acids. Similar values have been established for the SFAs group from crude commercial salmon oil [18]; interestingly, refined commercial salmon oil and Asian catfish oil also showed these three FA as the major ones, but a higher presence of stearic acid compared with myristic acid was detected in our study [16,18]. Among the MUFAs, the most abundant in the present study were 18:1 9c, 16:1 9c, 18:1 11c, and 20:1 11c. The total n-3 long chain polyunsaturated fatty acids (*n*-3 PUFAs) content was 1.14 times higher than the content of total *n*-6 PUFAs, showing values higher than those previously reported in different kinds of oils obtained crude commercial salmon, and rainbow trout (*Oncorhynchus mykiss*) [18,20]. The analysis of process variables shows a high recovery performance of RCSO concentrate when the urea:FA content ratio has the lowest value, and the value considered for the crystallization temperature is the highest. The results show a higher content of EPA and DHA when the total FA yield is lower, indicating that this experiment eliminated most of the SFAs and MUFAs from the starting oil, only leaving a small fraction of such acids in the urea non-complexed fraction. Similar values have been observed by other authors [17,21], who reported major yields with higher stirring speed, corroborating a significant effect of the stirring speed on the total FA yield in the urea complexation process ($p < 0.05$). Current results show an inverse relationship between the urea:FA content ratio and the crystallization temperature on the DHA content, which is a conclusion that has already been than those reported by several authors, whose employed different oils from marine origin such as seal blubber oil, tuna oil, Asian catfish oil, and a by-product of rainbow trout [13,14,16,17]. In this case, results similar for EPA content were found. As an explanation for this, it could be argued that the urea:FA content ratio has a significant positive effect ($p < 0.05$) on the concentration of EPA and DHA, probably as a result of increasing the concentration of urea with respect to that of FFA; consequently, this would lead to an increase of the number of adducts formed between the flat structures of SFA and urea molecules, this favoring the formation of hexagonal complexes and crystals [13], and leading to an increased concentration of EPA and DHA in the non-urea complexing solution. Previous studies on tuna oil concentrates have revealed that the regression models for total FA yield and total EPA+DHA content were highly significant with satisfactory R^2 coefficients [14,22]. The present data proved that the urea complexation under the optimization process conditions has shown to be highly efficient, since an increase in the total PUFAs content from 38.24 to 96.99 g/100 g total FA could be reached, which was 1.2 times higher than that reported by Pando et al. [18]. Additionally, the total EPA+DHA final content was 2.5 times higher than that reported for refined salmon oil concentrate without applying an optimization process [18]. On the other hand, the oxidative and hydrolytic stability parameters (PV, pAV, TOTOX, and FFA) indicate that the quality of the oil used in this study was included within the acceptable quality limits for edible fats and oils of marine origin according to the Chilean Food Sanitary Regulations. Government institutions and trade associations, such as the Council for Responsible Nutrition (CRN) and the Global Organization for EPA and DHA Omega-3s (GOED), have set strict guidelines for marine oil quality and safety parameters. According to the previously described recommended values [23,24], the starting RCSO can be considered included in the 98% of fish oil products compliant with the PV limit of 10 meq. active oxygen kg^{-1} oil set by British Pharmacopeia Fish Oil Type I, European Union (EU) Pharmacopeia Fish Oil Type I, and Australian government guidelines. Furthermore, salmon oil samples were compliant with the pAV limit of 15 set by British and EU Pharmacopeia Fish Oil Type II, as well as the pAV limit of 20 set by the GOED, Canada Natural Health Product Directorate (NHP), United States Pharmacopeia, and Codex Alimentarius Commission, Food and Agriculture Organization of the United Nations (CODEX/FAO). Thus, the primary and secondary lipid oxidation levels of RCSO are not exceeding the regulatory thresholds in the testing peroxide values and p-anisidine values. Interestingly, such lipid oxidation scores can be considered similar to those previously obtained for refined salmon oil samples [18] and salmon oil [24]. On the other hand, the FA composition of the RCSO showed that most abundant FAs were C18:1 9c, 18:2 9c, 12c, and C16:0, followed by EPA and DHA. The value of the total SFAs was found to be 21.28% in RCSO. Concerning PUFAs, the most abundant in RCSO concentrates were DHA, EPA, and 18:2 9c, 12c.

4. Materials and Methods

4.1. Materials and Chemicals

Refined commercial salmon oil was provided by Fiordo Austral S.A. (Puerto Montt, Chile). Fatty acid methyl ester (FAME) standards, fatty acid (FA) standards, and C23:0-methyl ester (2COT N-23M-A29-4 NU-CHECK-PREP-INC) were obtained from NU-CHECK-PREP, INC (Elysian, MN, USA). All the solvents and chemicals used (including urea, ethanol, α-tocopherol, and *n*-hexane) were of analytical grade (Merck, Santiago, Chile).

4.2. Characterization of Refined Commercial Salmon Oil

Initial RCSO characterization was carried out by chemical analyses. For it, the following standard Association of Official Analytical Chemists (AOAC) and official methods [25] were carried out: peroxide value (PV; method Cd 8b-90:1-2), p-anisidine value (pAV; method Cd 18-19:1-2), and total oxidation value (TOTOX; method Cg 3-91) and free fatty acids (FFA) contents (method Ca 5a-40:1).

4.3. FA Composition of the RCSO and n-3 LCPUFA Concentrates

For analyzing the FA composition of the RCSO and the different LCPUFA concentrates, a methylation process was performed to obtain FAMEs. For it, a two-step process was performed, according to previous research [18,26]. FAME analysis was carried out on an HP 5890 series II GLC with a flame ionization detector (FID) with the injection system split. A fused silica capillary column (100 m length × 0.25 mm × 0.2 μm film thickness) coated with SPTM-2560 (Supelco, Bellefonte, PA, USA) was used [18,26–29]. DataApex ClarityTM software (DataApex Ltd., Prague, Czech Republic) for chromatogram analysis was applied. The reference standard NU-CHEK GLC463 was used to identify the FA profiles. The concentration of the different FAME was determined from the calibration curves by assessment of the peak/area ratio. The quantification of all the individual FAs (g/100 g total FA) was achieved by employing C23:0 methyl ester as the internal standard according to the (AOCS Official Method (Ce 1j-7, 2009) [26].

4.4. n-3 LCPUFA Concentrates from RCSO

The procedure included salmon oil saponification, which was in agreement with other authors [17,18,30]. Concentrate from RCSO was prepared by FFA collection, the formation of urea FFA inclusion complexes, and the extraction of free n-3 LCPUFA. For it, the urea complexation method was carried out by 28 experimental runs with different urea:FA content ratios (0 to 6 *w/w*), crystallization temperatures (−30 to 30 °C), crystallization times (0 to 48 h), and stirring speeds (0 to 800 rpm). The FFAs were mixed with urea and 95 % ethanol, and the mixture was subsequently stirred and heated at 60 °C with magnetic stirring. Then, it was cooled with constant stirring to different conditions of temperature and time as described in the experimental design. The crystals formed were separated from the liquid phase by filtration with a Whatman No.1 paper. The non-urea complexing fraction was diluted with 100 mL of distilled water by each 10 g, acidified to pH 4.5 with 6 N HCl, and washed in a separating funnel with hexane (400 mL). The hexane phase was filtered with anhydrous sodium sulfate in a Whatman No.1 paper, and the solvent was partially removed using a rotatory evaporator at 40 °C under vacuum [16,17]. The resulting n-3 LCPUFA concentrates were stored at −80 °C with 0.5% of α-tocopherol under nitrogen atmosphere until use for further analysis.

4.5. Experimental Design and Optimization Procedure

A rotational central composite design 2^4 + star, with 4 factors and 5 levels, was carried out, with 28 experimental runs that included 4 repetitions of the central point based on the RSM. The following conditions for the independent variables were considered (Table 1): urea:FA content ratio (variable A: 0 to 6 w/w), crystallization temperature (variable B: −30 to 30 °C), crystallization time (variable C: 3.05 to 48.0 h), and stirring speed (variable D: 0 to 800 rpm). On the basis of the non-urea complexing fraction, the following response variables (R variables) of the experiment design were chosen: total FA yield (variable R1: g FA in the non-urea complexing fraction/100 g initial RCSO), EPA content (variable R2: g/100 g total FA in concentrate), and DHA content (variable R3: g/100 g total FA in concentrate). Four replicates were carried out at the central point of the experimental design in order to evaluate the experimental error. All experiments were performed randomly to minimize the effect of unexplained variability in responses resulting from extraneous factors [31]. In order to obtain response surfaces, multiple regression equations were fitted to the responses obtained by discarding non-significant terms ($p > 0.05$) to obtain response surfaces. To maximize the desirability function a multiple response optimization was performed to optimize several responses simultaneously, which maximized the desirability function scores that ranged between 0 and 1 [32]. The RSM was used to optimize and maximize the response variables, and a quadratic polynomial regression model was assumed for predicting individual Y variables. The model proposed for each Y value was as according to Equation (5):

$$Y_i = \beta_0 + \sum_{i=1}^{4} \beta_i X_i + \sum_{i=1}^{4} \beta_{ii} X_i^2 + \sum_{i=1}^{3} \sum_{j=i+1}^{4} \beta_{ij} X_i X_j + \varepsilon, \tag{5}$$

where β_0, β_i, and β_{ii} represent the intercept, linear, and quadratic coefficients, respectively; β_{ij} corresponds to the interaction coefficient terms for the interaction of variables i and j; X_i represents the independent variables; and ε denotes to the random error [31,32].

4.6. Statistical Analysis

Multiple regression analysis, ANOVA. canonical, and ridge maximum of data in the response surface regression (RSREG) procedure was used. The estimated response surface and contours of the estimated response surface were developed using the fitted quadratic polynomial equations obtained from the response surface regression (RSREG) analysis, holding the independent variables with the least effect on the response at a constant value and changing the levels of the other two variables [31,32]. A multiple-response optimization was performed to assess the combination of experimental factors that simultaneously optimize several responses; as a result, maximization of the desirability function was obtained, this function ranging from 0 to 1 [32]. Analyses were performed in triplicate considering the standard deviation of each sample. The lack-of-fit test was carried out by comparison of the variability of the current model residuals with the variability between observations at replicate settings of the factors [31,32]. Statgraphics Centurion XV.II (Manugistics Inc., Rockville, USA) was used.

5. Conclusions

The physical–chemical analyses of refined commercial salmon oil indicate that it is a good quality raw material that complies with the characteristics that are typical of oils of marine origin. A high concentration of EPA and DHA was obtained from RCSO, achieving an increase of up to 4.1 and 7.9 times in the concentrate, with values of 31.20 and 49.31 g/100 g total FA, respectively. Interestingly a 5.8-time increase was observed for the EPA+DHA content, from 13.78 to 80.51 g/100 g total FA. Therefore, it has been proved that it is possible to maximize the EPA and/or DHA content by the optimization of the variables in the urea inclusion process. The results of this study could serve as a basis for the food and nutraceutical industry, whose processes with clean technologies can develop new functional foods enriched with EPA and DHA. Furthermore, the consumption of such new functional foods would be likely to produce a positive and profitable impact on the health of a wide range of consumers such as pregnant women, infants, and older adults in general.

Author Contributions: Conceptualization, A.R., J.O.-V., S.P.A., and M.T.; methodology, E.C., A.R., M.M., and A.E.; data curation, M.M. and A.R.; writing—original draft preparation, A.R., G.D.-R., and J.O.-V.; writing—review and editing, A.R., J.O.-V., A.E., and G.D.-R.

Funding: This research was funded by the FONDECYT program (Government of Chile) throughout, grant number 1181774.

Conflicts of Interest: The authors declare no conflict of interest.

References

1. Food and Agriculture Organization/World Health Organization (FAO/WHO). *Fats and Fatty Acids in Human Nutrition*; Report of an Expert Consultation; FAO/WHO: Rome, Italy, 2010; 166p, ISBN 978-92-5-106733-8.
2. Schuncka, W.; Konkelb, A.; Fischerb, R.; Weylandtc, K. Therapeutic potential of omega-3 fatty acid-derived epoxyeicosanoids in cardiovascular and inflammatory diseases. *Pharm. Therap.* **2018**, *183*, 177–204. [CrossRef] [PubMed]
3. Siscovick, D.S.; Barringer, T.A.; Fretts, A.M.; Wu, J.H.Y.; Lichtenstein, A.H.; Costello, R.B.; Kris-Etherton, P.M.; Jacobson, T.A.; Engler, M.B.; Alger, H.M.; et al. Omega-3 polyunsaturated fatty acid (fish oil) supplementation and the prevention of clinical cardiovascular disease: A science advisory from the American Heart Association. American Heart Association Nutrition Committee of the Council on Lifestyle and Cardiometabolic Health; Council on Epidemiology and Prevention; Council on Cardiovascular Disease in the Young; Council on Cardiovascular and Stroke Nursing and Council on Clinical Cardiology. *Circulation* **2017**, *135*, e867–e884.
4. Weylandt, K.; Schmöcker, C.; Ostermann, A.; Kutzner, L.; Willenberg, I.; Kiesler, S.; Steinhagen-Thiessen, E.; Schebb, N.; Kassner, U. Activation of lipid mediator formation due to lipoprotein apheresis. *Nutrients* **2019**, *11*, 363. [CrossRef]
5. Swanson, S.; Block, R.; Mousa, S.A. Omega-3 fatty acids EPA and DHA: Health benefits throughout life. *Adv. Nutr.* **2012**, *3*, 1–7. [CrossRef]
6. Haast, R.; Kiliaan, A. Impact of fatty acids on brain circulation, structure and function. *Prost. Leuk. Ess. Fatty Ac. J.* **2015**, *92*, 3–14. [CrossRef] [PubMed]
7. Minihane, A.; Armah, C.; Miles, E.; Madden, J.; Clark, A.; Caslake, M.; Calder, P. Consumption of fish oil providing amounts of eicosapentaenoic acid and docosahexaenoic acid that can be obtained from the diet reduces blood pressure in adults with systolic hypertension: A retrospective analysis. *J. Nut.* **2016**, *146*, 516–523. [CrossRef]
8. WHO. *World Health Statistics: Monitoring health for the SDGs, Sustainable Development Goals*; WHO: Geneva, Switzerland, 2016.
9. Ofosu, F.K.; Daliri, E.B.M.; Lee, B.H.; Yu, X. Current trends and future perspectives on omega-3 fatty acids. *Res. J. Biol.* **2017**, *5*, 11–20.
10. Zanoaga, O.; Jurj, A.; Raduly, L.; Cojocneanu-Petric, R.; Fuentes-Mattei, E.; Wu, O.; Braicu, C.; Gherman, C.D.; Berindan-Neagoe, I. Implications of dietary ω-3 and ω-6 polyunsaturated fatty acids in breast cancer (Review). *Experim. Therap. Med.* **2018**, *15*, 1167–1176. [CrossRef]
11. Rubio-Rodríguez, N.; Beltrán, S.; Jaime, I.; De Diego, S.M.; Sanz, M.T.; Rovira, J. Production of omega-3 polyunsaturated fatty acid concentrates: A review. *Innov. Food Sci. Em. Technol.* **2010**, *11*, 1–12. [CrossRef]
12. Haq, M.; Getachew, A.T.; Saravana, P.S.; Cho, Y.J.; Park, S.K.; Kim, M.J.; Chun, B.-S. Effects of process parameters on EPA and DHA concentrate production from Atlantic salmon by-product oil: Optimization and characterization. *Kor. J. Chem. Eng.* **2016**, *34*, 2255–2264. [CrossRef]
13. Wanasundara, U.; Shahidi, F. Concentration of omega 3- polyunsaturated fatty acids of seal blubber oil by urea complexation: Optimization of reaction conditions. *Food Chem.* **1999**, *65*, 41–49. [CrossRef]
14. Liu, S.; Zhang, C.; Hong, P.; Ji, H. Concentration of docosahexaenoic acid (DHA) and eicosapentaenoic acid (EPA) of tuna oil by urea complexation: Optimization of process parameters. *J. Food Eng.* **2006**, *73*, 203–209. [CrossRef]
15. Patil, D. Recent trends in production of polyunsaturated fatty acids (PUFA) Concentrates. *J. Food Res Technol.* **2014**, *2*, 15–23.
16. Thammapat, P.; Siriamornpun, S.; Raviyan, P. Concentration of eicosapentaenoic acid (EPA) and docosahexaenoic acid (DHA) of Asian catfish oil by urea complexation: Optimization of reaction conditions. *J. Sci. Technol.* **2016**, *38*, 163–170.

17. Pando, M.; Rodríguez, A.; Galdames, A.; Berríos, M.M.; Rivera, M.; Romero, N.; Valenzuela, M.A.; Ortiz, J.; Aubourg, S.P. Maximization of the docosahexaenoic and eicosapentaenoic acids content in concentrates obtained from a by-product of rainbow trout (*Oncorhynchus mykiss*) processing. *Eur. Food Res. Technol.* **2018**, *244*, 937–948. [CrossRef]

18. Pando, M.; Bravo, B.; Berríos, M.M.; Galdames, A.; Rojas, C.; Romero, N.; Camilo, C.; Rodriguez, A.; Aubourg, S.P. Concentrating *n*-3 fatty acids from crude and refined commercial salmon oil. *Czech J. Food Sci.* **2014**, *32*, 169–176. [CrossRef]

19. Roca-Saavedra, P.; Mariño-Lorenzo, P.; Miranda, J.M.; Porto-Arias, J.J.; Lamas, A.; Vazquez, B.I.; Franco, C.M.; Cepeda, A. Phytanic acid consumption and human health, risks, benefits and future trends: A review. *Food Chem.* **2017**, *15*, 237–247. [CrossRef]

20. Haliloğlu, H.I.; Bayır, A.; Sirkecioğlu, A.N.; Aras, N.M.; Atamanalp, P. Comparison of fatty acid composition in some tissues of rainbow trout (*Oncorhynchus mykiss*) living in seawater and freshwater. *Food Chem.* **2004**, *86*, 55–59. [CrossRef]

21. Guil-Guerrero, J.L.; Belarbi, H. Purification process for cod liver oil polyunsatured fatty acids. *J. Am. Oil Chem. Soc.* **2001**, *78*, 477–484. [CrossRef]

22. Fei, Ch.; Salimon, J.; Said, M. Optimization of Urea Complexation by Box-Behnken Design. *Sains Malaysiana.* **2010**, *39*, 795–803.

23. De Boer, A.; Ismail, A.; Marshall, K.; Bannenberg, G.; Yan, K.L.; Rowe, W.J. Examination of marine and vegetable oil oxidation data from a multi-year, third-party database. *Food Chem.* **2018**, *254*, 249–255. [CrossRef]

24. Méndez, C.; Masson, L.; Jiménez, P. Estabilización de aceite de pescado por medio de antioxidantes naturales. *Aceites y Grasas* **2010**, *80*, 492–500.

25. AOAC. *Official Methods and Recommended Practices of the American Oil Chemists Society (edited by AOCS)*; AOAC: Champaign, IL, USA, 1993.

26. AOCS. *Determination of Cis-, Trans-, Saturated, Monounsaturated, and Polyunsaturated Fatty Acids by Capillary Gas Liquid Chromatography (GLC). Official Method Ce 1j-7. Official Methods and Recommended Practices of the American Oil Chemists Society (edited by AOCS)*; AOCS: Champaign, IL, USA, 2009.

27. Tran, Q.T.; Le, T.T.T.; Pham, M.Q.; Do, T.L.; Vu, M.H.; Nguyen, D.C.; Bach, L.G.; Bui, L.M.; Pham, Q.L. Fatty acid, lipid classes and phospholipid molecular species composition of the marine Clam *Meretrix lyrata* (Sowerby) from Cua Lo Beach, Nghe An Province, Vietnam. *Molecules* **2019**, *24*, 895. [CrossRef]

28. Kieliszek, M.; Błażejak, S.; Bzducha-Wróbel, A.; Kot, A.M. Effect of selenium on lipid and amino acid metabolism in yeast cells. *Biol. Trace Elem. Res.* **2019**, *187*, 316–327. [CrossRef]

29. Berríos, M.; Rodríguez, A.; Rivera, M.; Pando, M.; Valenzuela, M.A.; Aubourg, S.P. Optimization of rancidity stability in long-chain PUFA concentrates obtained from a rainbow trout (*Oncorhynchus mykiss*) by-product. *Int. J. Food Sci. Technol.* **2017**, *52*, 1463–1472. [CrossRef]

30. Zuta, C.P.; Simpson, B.K.; Chan, H.M.; Phillips, L. Concentrating PUFA from Mackerel Processing Waste. *J. AOCS.* **2003**, *80*, 933–936. [CrossRef]

31. Myers, R.H.; Montgomery, D.C. *Response Surface Methodology: Process and Product Optimization Using Design Experiments*; John Wiley & Sons, Inc.: New York, NY, USA, 1995.

32. Derringer, G.; Suich, R. Simultaneous Optimization of Several Response Variables. *J. Qual. Technol.* **1980**, *12*, 214–219. [CrossRef]

Sample Availability: Samples of the compounds are not available from the authors.

Review

Potential Use of Marine Seaweeds as Prebiotics: A Review

Aroa Lopez-Santamarina, Jose Manuel Miranda *, Alicia del Carmen Mondragon,
Alexandre Lamas, Alejandra Cardelle-Cobas, Carlos Manuel Franco and Alberto Cepeda

Laboratorio de Higiene Inspección y Control de Alimentos, Departamento de Química Analítica, Nutrición y
Bromatología, Universidade de Santiago de Compostela, 27002 Lugo, Spain; a_lo_san@hotmail.com (A.L.-S.);
aliciamondragon@yahoo.com (A.d.C.M.); alexandre.lamas@usc.es (A.L.); alejandra.cardelle@usc.es (A.C.-C.);
carlos.franco@usc.es (C.M.F.); alberto.cepeda@usc.es (A.C.)
* Correspondence: josemanuel.miranda@usc.es; Tel.: +34-982-252-231 (ext. 22407); Fax: +34-982-254-592

Academic Editors: Sylvia Colliec-Jouault and Jesus Simal-Gandara
Received: 16 December 2019; Accepted: 21 February 2020; Published: 24 February 2020

Abstract: Human gut microbiota plays an important role in several metabolic processes and human diseases. Various dietary factors, including complex carbohydrates, such as polysaccharides, provide abundant nutrients and substrates for microbial metabolism in the gut, affecting the members and their functionality. Nowadays, the main sources of complex carbohydrates destined for human consumption are terrestrial plants. However, fresh water is an increasingly scarce commodity and world agricultural productivity is in a persistent decline, thus demanding the exploration of other sources of complex carbohydrates. As an interesting option, marine seaweeds show rapid growth and do not require arable land, fresh water or fertilizers. The present review offers an objective perspective of the current knowledge surrounding the impacts of seaweeds and their derived polysaccharides on the human microbiome and the profound need for more in-depth investigations into this topic. Animal experiments and in vitro colonic-simulating trials investigating the effects of seaweed ingestion on human gut microbiota are discussed.

Keywords: prebiotic; polysaccharides; seaweed; Rhodophyceae; Phaeophyceae; Chlorophyceae

---

1. Introduction

Marine seaweeds have been consumed whole by East Asian populations for centuries, if not millennia, appearing in traditional recipe books in many countries [1]. Additionally, their human consumption in Western countries has been increasing in the latest decades because of their association with improved human health. Some benefits of their consumption include a lower incidence of cancers, decreased blood pressure and blood sugar, and antiviral, anti-inflammatory, immunomodulatory or neuroprotective activities [1,2]. A mechanistic link proposed to explain the prevention of such diseases by seaweed consumption implicates the presence of diverse health-promoting bioactive compounds in seaweeds, including sulphated polysaccharides, polyphenols, pigments (chlorophylls, fucoxanthins, phycobilins), carotenoids, omega-3 fatty acids or mycosporine-like amino acids [1,3,4]. In some cases, these compounds are not produced by terrestrial plants or are not consumed in adequate quantities as part of a typical Western diet.

Nowadays, with a continuously expanding world population, fresh water is an increasingly scarce commodity, and the world's desertification processes continue. About 45% of the world's land surface is considered drylands, while 12 million hectares of land are degraded yearly through lack of water and related processes [5]. According to the Food and Agriculture Organization of the United Nations [6], agricultural productivity is persistently declining at over 1% per year. It is thus reasonable to expect that in the next decades there will be a need for increasing algae production to replace or supplement the intake of plant foods of terrestrial origin. Seaweeds have numerous advantages over terrestrial plants, such as rapid growth rates, and they do not require arable land, fresh water or contaminating fertilizers [7]. Additionally, an increase in seaweed cultivation would provide environmental services as an added benefit [5].

Marine seaweeds constitute approximately 25,000–30,000 different species, with a great diversity of forms and sizes [8]. The taxonomic groups, which reflect their pigmentation, include red algae (Rhodophyceae), brown algae (Phaeophyceae) and green algae [8]. Although they are still produced on a very modest scale relative to global food production, their worldwide cultivation has increased rapidly in the last decades, reaching a current yearly production of about 29 million tonnes [8]. Asian countries dominate world production, with 99% of the total, while most other maritime countries produce little or none [5]. Polysaccharides account for the majority of seaweed biomass (up to 76% of dry weight in some species; [8] and together with oligosaccharides have been the key focus of many studies of seaweed-derived compounds. Phenolic compounds and proteins from seaweeds have also attracted interest as potential functional ingredients [8]. Besides their consumption as an entire food, seaweeds or their polysaccharides are considered valuable additives in the food industry because of their rheological properties as gelling and thickening agents [2]. Additionally, seaweeds were largely employed in the formulation of animal feed [9], as well as in the formulation of cosmetics, drugs and fertilizers [10].

Regarding specific benefits on human health, seaweeds have demonstrated to exert preventive effects against several non-transmissible diseases such as cardiovascular diseases [11,12], antihypertensive [13], anti-obesity effects [14] and anti-diabetic effects [15,16], anti-cancer [17,18] or antioxidant activities [19].

Regarding cardiovascular diseases, there are large contributing risk factors that overlap and intertwine, contributing overall to the onset and growth of the disease [11]. Between them, it can be cited a cascade of mechanisms including vascular inflammation, oxidative stress, hypercoagulability and activation of the sympathetic and renin-angiotensin systems. Although in some cases the exact mechanisms through seaweeds can prevent cardiovascular diseases are not always fully understood, it was demonstrated that seaweed consumption can prevent cardiovascular diseases [11,12]. Functional oligosaccharides from seaweeds are associated with a variety of biological processes linked to hypoglycaemic and hypolipidaemic activities, although the concrete mechanisms have not been well studied [15]. With respect to hypertension, various compounds from seaweeds, such as protein-derived bioactive peptides and phlorotannins, can prevent hypertension by inhibition of angiotensin-I converting enzyme activity [13].

The potential anti-obesity activity derived from seaweeds consumption may involve a large variety of mechanisms and alterations in lipid metabolism, suppression of inflammation, suppression of adipocyte differentiation and delay in gastric emptying [14]. Between then, an important anti-obesity activity of seaweeds is the inhibition of peroxisome proliferator-activated receptor γ (PPARγ) expression and activation of the adenosine monophosphate-activated protein kinase (AMPK) phosphorylation [13]. Other important anti-obesity mechanism of seaweeds is related with inhibition of lipases, especially pancreatic lipase, that is one of the main therapeutic targets of anti-obesity drugs [14] and was recently demonstrated for various seaweeds species [20]. Additionally, anti-obesity mechanisms related with concrete seaweed components, such as phlorotannins, that target the inhibition of adipocyte differentiation or fucosterol, that decreases the expression of the adipocyte marker proteins PPARγ and CCAAT/enhancer-binding protein alpha were reported. Seaweeds can also prevent obesity by means of the modification of the relative quantities of phyla in the gut microbiota (GM), and polysaccharides from seaweeds can reduce obesity also by repairing the intestinal barrier and reducing inflammation [21].

Although abundant signaling pathways have been found to be involved in the process of glucose metabolism and anti-diabetic effects [15,16], among them, the IRS1/PI3K/JNK/AKT/GLUT4 pathways are important mechanisms in insulin signal transduction. Some seaweed components were shown in animal models to ameliorate the hepatic insulin resistance by regulating the cited signaling pathways [15,16,22]. In addition, seaweeds significantly increased the abundance and diversity of gut microbiota in animal models and showed the ability to increase the population of beneficial microbiota and maintain the homeostasis of the GM [15,22].

With respect to the antioxidant activity of seaweeds, this protective affects depends mainly on phlorotannins, secondary metabolites that exert their great antioxidant activity via the scavenging of reactive oxygen species [19]. Secondary metabolites of seaweeds are also responsible for the anti-cancer activity, whose includes different mechanisms such as repairing the intestinal barrier by intensifying the expression of the tight junction proteins via increasing the phosphorylation of MAPK and ERKT/2 genes [18], and activating the caspase cascades. Other potential mechanisms are reducing the expression of cyclin-dependent kinases and matrix metalloprotease family [17] and inducing decreased levels of pro-apoptotic metabolic signals [19].

2. Polysaccharides from Marine Seaweeds

Indigestible dietary polysaccharides attract attention as functional food ingredients with health benefits [7]. Most carbohydrates entering the colon are fermented in the proximal colon, which is considered a saccharolytic environment. As digesta moves through the distal colon, carbohydrate availability decreases, and proteins and amino acids become the main metabolic energy sources for bacteria in this region. The main end-products of saccharolytic fermentation are short chain fatty acids (SCFA), which contribute towards the host's daily energy requirements. On the contrary, the end-products of proteolytic fermentation include metabolites such as phenolic compounds, and nitrogenous ones like amines and ammonia, some of which are carcinogens. This means that the GM exerts a key contribution to the human energy balance and nutrition, by extending the host metabolic capacity to indigestible polysaccharides. In addition, intestinal microorganisms contribute to develop and maintain the host immune system, defending the host from colonization by opportunistic pathogens [2]. The effects of polysaccharides on the GM are generally evaluated by the contents of SCFAs, the composition and the abundance of beneficial intestinal bacteria [20].

Polysaccharides in today's human diet originate primarily from terrestrial plant cell walls, while other sources, such as seaweeds, are less represented [8]. Studies indicate that polysaccharides and oligosaccharides derived from seaweeds can modulate intestinal metabolism, including fermentation, inhibit pathogen adhesion and evasion, and potentially treat inflammatory bowel disease [8,23]. Some seaweed polysaccharides also demonstrate anticoagulant [24], antitumor [25], anti-inflammatory [26], antiviral, antihyperlipidemic [27] or antioxidant activity [28]. Other research has focused on their use as prebiotics to aid in limiting the occurrence of non-transmissible chronic diseases common in Western countries, such as obesity, diabetes, cardiovascular diseases or some types of cancer [2]. Nevertheless, many seaweed fibres are high-molecular-weight polymers that need to be transformed into oligosaccharides to increase their fermentability by the GM [2].

Depending on the taxonomic classification of algae, polysaccharides can vary greatly in their composition [8]. Seaweeds feature an integrated network of biopolymers in their cell walls, mainly formed by polysaccharides associated with other compounds, such as proteins, proteoglycans, polyphenols and some mineral elements, like calcium and potassium [8]. It is because of this complexity that for most seaweed polysaccharides, the exact structures, constituents and chemistry are not fully known [29]. Depending on the algal taxa, both structural and storage polysaccharides may vary. Structural polysaccharides are the most abundant, and their composition can be influenced by the seaweed species [29], as well as environmental factors, such as salinity, water temperature and sunlight intensity [30]. Some of the structural polysaccharides are carboxylated or sulphated, which can affect their fermentability [2].

Green seaweeds contain mostly sulphated structural polysaccharides, like ulvans (the most abundant, representing 8–29% of dry weight) and sulphated galactans, xylans and mannans. These polymers are composed mainly of rhamnose, xylose, glucose, glucuronic acid and sulphates, with smaller amounts of mannose, arabinose and galactose [31,32]. These polysaccharides are not fully fermented by the human GM [33,34]. Conversely, the main carbohydrate in storage is starch.

Contrariwise, brown seaweeds contain mainly cellulose, alginic acids, fucoidans and sargassans as structural polysaccharides, while the storage polysaccharides are alginates [35] (the most abundant at 17–45% of dry weight), fucoidans and laminarins [2,8,36]. Finally, red seaweeds contain agars, carrageenans, xylans, sulphated galactans and porphyrins as main structural polysaccharides, while the main storage polysaccharide is starch [21,37].

Through a long time of co-evolution between GM and host, the intestinal microbes have evolved diverse strategies for degrading polysaccharides from terrestrial plants [38]. However, because the consumption of seaweeds polysaccharides was not common over human evolution, the human GM did not acquire the same efficacy to degrade seaweed polysaccharides. Thus, although humans possess the enzymes necessaries to degrade some algal polysaccharides, such as starches, they are unable to digest the most complex polysaccharides [2,38]. An elegant work carried out by Hehemann et al. [39], showed that specific genes coding for enzymes with potential capacity to degrade seaweed polysaccharides, as porphyranases or agarases, can be transferred from a member of marine Bacteroidetes, *Zobellia galactanivorans* to the GM bacterium *Bacteroides plebeius* in Japanese individuals. As a consequence, GM of those subjects acquires the ability to degrade porphyran and agarose, as compared to the GM from North American individuals, who are incapable of degrading it [39].

De Jesus Raposo et al. [40] suggested that most seaweed polysaccharides can be regarded as dietary fibre, as they are resistant to digestion by enzymes present in the human gastrointestinal tract, reaching the distal gut. In this allocation, polysaccharides are fermented and become food for the commensal bacteria, stimulating their growth. For this reason, great efforts have been placed on developing efficient methods for seaweed polysaccharides extraction, purification and structural characteristics elucidation in order to improve their bioavailability, especially for insoluble fibre [2].

It was previously shown that variations in the chemical structure of a prebiotic can impact its selective fermentation by bacteria [41]. For this reason, there were published large works regarding the investigation of the potential prebiotic effect of single polysaccharides, often contained in seaweeds, but employing pure standards [42–44]. However, it should be considered that seaweeds contain other components than also can affect GM. Consequently, trials employing whole seaweeds are required to investigate their real effect in human GM. Furthermore, although several works investigated the effects of seaweeds in livestock gut microbiota, they are more oriented to study the effects of seaweeds in animal production and welfare. An elegant review was recently published where they can be checked [32].

3. Other Bioactive Compounds from Marine Seaweeds

In addition to polysaccharides, seaweeds also contain other bioactive compounds, called secondary metabolites much of them with antioxidant activity [45]. Among these, polyoletides (such as phlorotannins), isoprenoids (such as terpenes, carotenoids and steroids), alkaloids and shkimates (such as flavonoids) are the main groups of secondary metabolites found in algae [46]. Compared with other macroalgae, red seaweeds are richer sources of these secondary metabolites [47]. The human health benefits afforded by these bioactive compounds include anti-inflammatory, antioxidant, anticoagulant, antiviral, antimicrobial, antidiabetic, antitumor, antihypertensive, antiallergic and immunomodulatory activities [45–48].

Exceptionally, phlorotannins or polyphenols are recognized as structural classes of polyketides, found primarily in brown algae. These compounds can also reach the large intestine where GM can convert them into beneficial bioactive metabolites. Phlorotannins are highly hydrophilic components formed by polymerization of monomeric units of phloroglucinol (1,3,5-trihydroxybenzene). There are six main groups: fucols, floretoles, fucofloretols, fuhalols, isofuhalols and eckols, and all display strong antioxidant properties and act against oxidative stress [46]. Certain polyphenols are used as prophylactics against problems such as cardiovascular diseases, cancers, arthritis and autoimmune disorders [47]. In addition, some phlorotannins have been shown to decrease blood glucose levels after carbohydrate-rich meals. This action is achieved by interfering with the enzymes amylase and sucrase that intervene in the digestion and assimilation of these carbohydrates. In addition to their effects on the metabolic functions of the host, phlorotannins also appear to have some antibacterial activities [47], which may explain the low production of AGCC derived from seaweeds in which they are an important part [7]. However, like other phenolic compounds, bacterial growth inhibition occurs selectively in microbial populations, including some pathogens, and its antibacterial effect is minor in commensal bacteria [7]. Because of this selective inhibition of bacterial pathogens, large whole seaweeds and seaweeds ethanolic extracts has been used to extend the shelf life of fresh fishery foods, such as *Fucus spiralis* [48,49], *Bifurcaria bifurcata* [50], *Cytoseira compressa* [51] or *Gracilaria verrucosa* [52] Another of its potentially therapeutic functions is that extracts with high phlorotannins content have demonstrated a potent inhibitory action on the growth of cancerous cell lines [53,54].

Bromophenols present in marine algae have attracted much attention in the field of antimicrobial agents [55]. Previous studies indicate that marine bromophenols possess promising antibacterial [56,57] and antiviral activities [58]. In addition, symphyocladin G, a new bromophenol adduct derived from the red seaweed *Symphyocladia latiuscula*, is found to have antifungal activity against *Candida albicans* [59]. Several bromophenols isolated from the red alga *Odonthalia corymbifera*, are promising candidates for antifungal agents in crop protection [56]. These properties are not exclusive to red algae because compounds, such as bis(2,3-dibromo-4,5-dihydroxybenzyl) ether, isolated from brown algae *Leathesia nana*, showed cytotoxic activity against some cancer cells [54], and exhibited antibacterial activity against several strains of Gram-positive and Gram-negative bacteria [56].

With respect to terpenes, there are about 200 different diterpenoids, of which some have important cytotoxic and antiviral, antimicrobial and antiparasitic activities (such as against *Leishmania*) [60,61]. These compounds are found in red and brown algae.

Flavonoids and their glycosides are present in green, brown and red algae. These compounds possess antioxidant properties and have demonstrated action against arteriosclerosis and cancer [45]. Within this group, fucoxanthin, β-carotene and violaxanthin stand out. Besides its strong anticancer activity, fucoxanthin has promise in preventing obesity [62]. The correlation between a carotenoid-rich diet and a low risk of cardiovascular and ophthalmological diseases has been supported by recent research with different types of carotenoids in cellular systems and human intervention studies [63]. Specifically, flavonoids from *Enteromorpha prolifera* influenced the GM balance in diabetic mice, increasing the presence of *Alistipes*, *Lachnospiraceae* and *Odoribacter* genera [63]. *Alistipes* spp. is one of the most abundant bacterial genera in the mouse intestine and is capable of fermenting glucose and lactic acid to produce propionic, acetic and succinic acid, which modulate the release of intestinal hormones, thereby influencing the release of insulin and appetite. It is perhaps for this reason that *E. prolifera* is traditionally used in China as a natural herb to treat diseases associated with inflammation [63]. It has recently been observed that a polysaccharide of *E. prolifera* could be used as a novel agent to treat obesity and hyperlipidaemia [62].

Other important secondary metabolites contained in seaweeds and responsible of important beneficial effect in human health are peptides, such as lectins [48]. Lectins primarily show antiviral, antibacterial, and antifungal activities. Specially one type of lectin, griffithsin, showed important antiviral activity and is nowadays considered a promise antiviral agent, with great potential concerning the prevention of sexually transmitted infections [48], including HIV [64]. Other important peptides are renin inhibitor tridecapeptide [65] and dipeptide [66], which demonstrated hypotensive effect dipeptide. Phycoerythrin [43] and kahalalide F [67] are other important peptidic compounds isolated from seaweeds than showed antitumor effect.

Besides the so-called secondary metabolites, seaweeds contain other minor nutrients of immense importance for human health. Phycobiliproteins, responsible for the characteristic bright pink appearance of red algae, are classified into phycoerythrin (red) and phycocyanin (blue). These pigments are used commercially in food, nutraceuticals, and for their therapeutic properties, mainly antimicrobial, antioxidant, anti-inflammatory, neuroprotective, hepatoprotective, immuno-modulatory and anticancer effects [67–72]. Such compounds may improve the efficacy of standard anticancer drugs, decrease their side effects, and act as photosensitisers for the treatment of tumour cells [30].

4. Effects of Seaweed Polysaccharides on Human Health

Compounds with prebiotic activity, such as oligosaccharides, lactulose, fructo-oligosaccharides (FOS), inulin, galacto-oligosaccharides and arabinoxylano-saccharides are used as functional ingredients in the food industry [10]. While most of the above compounds are now derived from terrestrial plants, some studies have shown that polysaccharides and oligosaccharides derived from marine algae can also modulate intestinal metabolism, including fermentation, inhibit adhesion and invasion of pathogens, and treat inflammatory bowel disease [23,73]. Furthermore, these compounds have demonstrated anticoagulant, antioxidant, immunomodulatory, antitumor and antiviral activities [10].

Being much less degradable by enzymes from the human upper gastrointestinal tract than their terrestrial plant counterparts, polysaccharides from marine algae reach a greater proportion in the descending colon. For this reason, some authors [42,74] have found that polysaccharides from marine algae, such as alginate, agarose oligosaccharides and κ-carrageenan oligosaccharides, have a higher prebiotic activity than FOS in vitro. Specifically, sulphated polysaccharides from marine algae show anticoagulant, antiviral, antitumor, anti-inflammatory, antibacterial, immunological, antioxidant and many other biological and physiological activities [8,75]. Sulphated polysaccharides include fucoidans (L-fucose and sulphated ester groups) from brown seaweeds, agars and carrageenans (sulphated galactans) from red seaweeds, and ulvans (sulphated glucuronoxylorhamnan) and other sulphated glycans from green seaweeds [8].

The consumption of these sulphated polysaccharides can block the adhesion of leukocytes to the epithelium of blood vessels, preventing the migration of these cells to the site of inflammation [76]. These polysaccharides often stimulate the growth and activity of beneficial bacteria by acting as substrates for fermentation in the large intestine, leading to the production of SCFA, with multiple functions that help maintain health [8]. As previously mentioned, seaweed polysaccharides differ in their properties and compositions from one type of algae to another [46], so their effects on the human GM will also differ.

The GM, especially in its most distal parts, harbors many bacteria, archaea, protozoa and viruses, which along with their genetic material, is collectively referred to as the gut microbiome (GMB) [77]. This GM is composed of up to 12 different bacterial phyla of which more than 90% belong to the Proteobacteria, Firmicutes, Actinobacteria and Bacteroidetes [78], while the remaining phyla are much less constant and numerous [79]. The most frequent bacterial species in the colon, which is where the highest bacterial concentration exists [78], belong mainly to the families *Bacteroidaceae*, *Prevotellaceae*, *Rikenellaceae*, *Lachnospiraceae* and *Ruminococcaceae* [77]. The GM presents a diverse set of functions important to human health, such as the extraction of energy from a broad spectrum of nutrients, the production of vitamins, the promotion of immune homeostasis and the prevention of colonization of the intestine by pathogens [80]. One of the most important functions of the GM is in the prevention of chronic low-grade inflammation [81]. Host genetics define the chemistry and physics of the GM, including the availability of nutrients and the threshold of activity required to induce an immune response. Consequently, intestinal microbial communities are composed of species that have evolved to occupy specific ecological niches in the gut, including the ability to metabolize specific molecules available from the host or to evade host defenses [77].

Nutrients can interact directly with the GM to promote or inhibit its growth. In this sense, the ability of the GM to extract energy from specific components of the diet offers a direct competitive advantage to specific members of the GM, allowing them to proliferate at the expense of other members [81]. Thus, diet affects not only the composition and absolute abundance of intestinal bacteria but also their growth kinetics [82]. In this context, the most influential nutrients are indigestible carbohydrates, which can be of both terrestrial and marine algae origin [81].

The human genome encodes a limited number of hydrolases capable of hydrolyzing the glycosidic bonds of polysaccharides in dietary fibre (collectively referred to as CAZymes). Consequently, many polysaccharides, such as resistant starch, inulin, lignin, pectin, cellulose and FOS, reach the large intestine undigested. In contrast, the GMB codes tens of thousands of CAZymes. In the presence of bacteria harboring key enzymes involved in carbohydrate metabolism, these complex polysaccharides can thus be degraded and metabolized in vivo [83]. The bacteria able to degrade these complex polysaccharides are called primary degraders and include members of the genera *Bacteroides*, *Bifidobacterium* and *Ruminococcus, Roseburia, Facealibacterium, Anaerostides* or *Coprococcus*. A relative abundance of these genera in our GM infers that during a food shortage, these bacteria can alternate between energy sources by using sensors and regulatory mechanisms that control gene expression [21, 81]. Hydrolases act on polysaccharides to generate oligosaccharides and monosaccharides. Secondary fermentation of these compounds by the GM produces SCFA, specifically acetic, propionic, butyric, lactic and succinic acids, which initiate a complex metabolic network [81].

The GM of hunter–gatherer, rural and agricultural populations are usually more bacterially diverse than in modernized urban societies [84] and so require a greater functional repertoire to maximize their energy intake from dietary fibres. Conversely, the consumption of a diet composed mainly of products of animal origin causes an enrichment in the GM of genera of bile-tolerant bacteria, such as *Alistipes*, *Bilophila* and *Bacteroides*, and the almost total exhaustion of bacteria that metabolize polysaccharides, such as *Roseburia, Eubacterium rectale* subgroup and *Ruminococcus bromii* [81].

Clinical studies investigating prebiotic effects have some disadvantages with respect to ethical constraints, as well as limited sampling possibilities from the colon and limited measurements of in situ SCFA production. These concerns are commonly avoided by applying an in vivo approach [41], that are the most common in the investigation of seaweed effects on human GM, as is described below. Contrariwise, in vitro studies show important limitations because only represent the first step of a long process, and the results observed in vitro can be magnified, diminished, or totally different in a more complex and integrated system [48]. An additional limitation is that, due to the short fermentation time in in vitro studies, they fails to capture the complete picture of cross-feeding interactions between gut microbes, and which may not fully correlate with the long-term effects of seaweed compounds on GM [41].

4.1. Polysaccharides from Brown Seaweeds

Although the prebiotic and immuno-modulation properties of brown algae have been studied both in animal models and in vitro, humans intervention studies are also needed to assess whether there is a direct association between these uses of algae and the human GM, but are currently restricted due to ethical concerns [78]. The most relevant results obtained from examining the impacts of brown seaweeds on the GM can be found in Table 1. In this table it were included results about the prebiotic effect of brown seaweed species from genus *Ecklonia* [7,85], *Sargassum* [86–88], *Laminaria* [82,89–92], *Ascophyllum* [93–95], *Fucus* [23,63], *Undaria* [90], *Saccorhiza* [96] or *Porphyra* [97]. As can be seem in Table 1, in most cases, the administration of whole brown seaweed or brown seaweed-extracted polysaccharides resulted in an increase of SCFA production, stimulating of beneficial bacteria grown such as *Lactobacillus* [7,82,85,86,95], *Bifidobacterium* [7,82,85,87,92] or *Faecalibacterium* [7,58,87]. In some cases, the brown seaweed or brown seaweed-extracted polysaccharides also inhibited the growth of potentially pathogen bacteria [73,86]. In some cases, it were reported other beneficial effects not strictly related with action on GM, such as reducing serum inflammatory markers [23], reducing serum levels of lipopolysaccharide-binding protein [44], increasing CAZymes [44], reducing activity of fecal bile salt hydrolase activity [96], or reduced the expression or diabetes-related genes [15].

Laminar storage polysaccharides, typical of brown seaweeds, are low-molecular-weight, linear polysaccharides composed of glucose units with a low degree of branching [79]. Besides affecting mucin composition and SCFA concentration, laminins can affect the adherence, translocation and proliferation of bacteria in the gut [98,99]. At the same time, laminins stimulate the proportion of *Bifidobacterium*, which generates a prebiotic potential. In other research, laminarin has been shown to promote an immune response [98], and could be useful for inhibiting the production of putrefactive substances from undigested proteins [100]. In vitro batch fermentation of laminarin for 24 h promoted an increase in *Bifidobacterium* and *Bacteroides*, and propionate and butyrate production [42]. Contradicting results by other researchers indicated that laminarin was not selectively fermented by *Lactobacillus* and

Bifidobacterium, but could modify the composition, secretion and metabolism of the jejunal, ileal, caecal and colonic mucosa to protect against bacterial translocation [32]. In addition, laminarin increased the presence of *Clostridium* spp. and *Parabacteroides distasonis* in rats [101].

An in vitro study conducted with the species *Sargassum thunbergii* revealed a dramatic increase in the population of beneficial bacteria (from 17% to 28%), while a group of harmful Firmicutes decreased from 75% to 64% after 48 h of fermentation [87]. No noticeable changes were found in Proteobacteria or Actinobacteria. At the genus level, an increase in *Lactobacillus*, *Bifidobacterium*, *Roseburia*, *Parasutterella* and *Fusicatenibacter* appeared after incubation for 24 h, followed by an increase in *Faecalibacterium* and *Coprococcus* at 48 h of incubation [87]. *Bifidobacterium*, *Coprococcus* and *Parasutterella* have been negatively correlated with non-alcoholic steatohepatitis, hepatocellular carcinoma and diabetes [102], while *Ruminococcus*, *Roseburia* and *Faecalibacterium* are producers of butyric acid and are facilitate the degradation of polysaccharides and fibres [103]. *Fusicatenibacter* was positively associated with increased serum leptin in obese rats [104], which reduces their appetite. All these findings highlight the prebiotic potential of *S. thunbergii* by its modulation of the composition and abundance of beneficial GM.

An in vitro study using *S. wightii* in MRS broth evaluated their antioxidant activity and prebiotic score comparing *L. plantarum* and *Salmonella* Typhimurium relative growths. The study showed that the prebiotic activity score was positive, promoting selectively the growth of *L. plantarum* with respect to the pathogen *S. Typhimurium*. Specifically, a prebiotic effect by 1.42-fold more growth stimulation of *L. plantarum* than *S. Typhymurium* [86].

In other work Chen et al. [58] showed an increase in fucoidan from *A. nodosum* in an in vitro assay simulating the human digestive tract was due to an increase in Bacteroidetes, Firmicutes and SCFA. At the genus level, the genera *Bacteroides*, *Phascolarctobacterium*, *Oscillospira* and *Faecalibacterium* increased, while the levels of *Fusobacterium*, *Megamonas*, *Parabacteroides*, *Clostridium* and *Dorea* decreased relative to the samples to which the algae *A. nodosum* had not been added [46]. demonstrated the in vitro prebiotic activity of a mixture of fucoidans and alginates obtained from *A. nodosum*, leading to an increase in the growth rate of *L. delbrueckii* and *L. casei* to levels similar to those observed after administration of inulin, a standard commercial prebiotic [46]. Other authors [93] conducted a study in rats, which were administered polysaccharides extracted from *A. nodosum*, and they were seen an increase in both acetate, propionate and butyrate SCFAs.

According to Zaporozhets et al. [76], fucoidans obtained from *F. evanescens* stimulate the colonic growth of beneficial *Bifidobacterium* species, such as *B. longum* B379M and *B. bifidum* 791B. Lean et al. [23] administered *F. vesiculosus*-derived fucoidan extracts to mice and, interestingly, found a reduction in markers associated with inflammatory bowel diseases.

When Wister rats were fed with feed enriched with alginates or laminarins, An et al. [101] found a notable decrease in the number of metabolites resulting from putrefaction, such as indole, H_2S and phenol. This result was subsequently confirmed in both in vitro and rat models by Nakata et al. [100], who also found a decrease in ammonium levels with alginate. At the phyla level, alginate increased the levels of Actinobacteria, while laminarins increased the levels of Proteobacteria. At the genus level, *Bacteroides* was markedly more abundant in the group fed with alginate, and *B. capillosus* was the most frequent species. In rats fed with laminarin-enriched feed, *Parabacteroides*, *Lachnospiraceae* and *Parasutterella* bacterium were detected in greater abundance than in control rats. Nguyen et al. [44] studied laminarin supplementation in a mice high-fat diet. They could see a decrease in Firmicutes and an increase in the Bacteroidetes phylum, especially the genus *Bacteroides*.

Ramnani et al. [94] performed in vitro fermentation with *A. nodosum*-derived alginates, which increased *Bifidobacterium* and SCFAs. An increase in the proportion of Bacteroidetes to Firmicutes was observed as well in fermentations added with sulphated polysaccharides extracted from *A. nodosum* versus controls. Increased levels of Bacteroidetes and decreased levels of Firmicutes have been associated with a reduced risk of obesity in humans [79].

Table 1. Prebiotic effect of different species of brown seaweed.

Type of Study	Seaweed, Dosage and Time of Exposure	Polysaccharides Characterization	Significant Changes in Gut Microbiota	Significant Changes in Related Metabolites	Reference
In vitro fermentation system using fresh fecal samples from four healthy donors	Polysaccharides extracted from 20 g of *Ascophyllum nodosum* in a single dose, compared to blank and FOS-added samples	Total carbohydrate 42.3%; uronic acid 11%; protein 1.4% and sulfate content 23.9%; Monosaccharides content were composed of Man, GlcA, Glc, Gal, Xyl, and Fuc at a molar ratio of 16.65, 20.34, 1.60, 9.69, 3.44, and 48.29	Increase in Bacteroidetes and Firmicutes. At genus level, increase of *Bacteroides*, *Phascolarctobacterium*, *Oscillospira*, *Faecalibacterium*, while decreased *Fusobacterium*, *Megamonas*, *Parabacteroides*, *Clostridium*, *Dorea*	Increase in SCFA, acetate and propionate in *A. nodosum* added polysaccharides with respect to blank samples and FOS-added samples	[58]
In vivo trials using 3 Wistar male rats per sample, comparing effect of *A. nodosum* crude polysaccharide with hydrolyzed *A. nodosum* polysaccharides. SCFA were generated by fermentation with *Lactobacillus plantarum* BCC 5493 and *Enterococcus faecalis* BCC 39,179	0.2 g of polysaccharides extracted from *A. nodosum* per rat for 4 days, comparing crude polysaccharide with crude polysaccharide hydrolysates, alginate and hydrolyzed alginate	Crude polysaccharide contained carbohydrates 22.7%, sulphate content of 17.1% and protein content 1.34%. Hydrolysates showed 25.1–26.7% carbohydrates, 25.3–25% sulphate and 1.7–1.4% protein contents	Not provided	Increase in both acetic, propionic and butyric acids, in this order. SCFA were higher in the case of polysaccharides with lower molecular weight	[93]
In vitro fermentation system using fresh fecal samples from three healthy donors	1% *w/v* low molecular weight polysaccharide derivatives extracted from *A. nodosum* for 24 h. Inulin was used as positive control and cellulose as negative control	Average molecular weight 31.0 and 56.0 kDa	No significant changes in GM	Increase in total SCFA, acetic and propionic acids	[94]
In vivo trial using 18 male C57BL/6 mice. 6 mice received fucoidans extracted fro *A. nodosum*, and 6 acted as blank group	100 mg/kg/day of fucoidans obtained from *A. nodosum* for 6 weeks. Control group received saline solution	21% sulfate content; 1330 KDa molecular weight; 7.3% Man, 24.1% GlcA, 1.5% Glc, 7.2% Gal, 1.3% Xyl, 58.6% Fuc	Fucoidans administration resulted in a much more diverse cecal microbiota, increase on *Lactobacillus* and *Talassospira*, whereas	Fucoidans decreased the serum levels of lipopolysaccharide-binding protein	[95]
In vitro fermentation system using fresh fecal samples from 3 healthy donors	1.5% *w/v* of enzyme-assisted extracted polysaccharides from *Ecklonia radiata* for 24 h. Inulin and resistant starch were used as positive controls and glucose and cellulose were used as negative controls	48.7% total fibre, 16.1% non-digestible non-starch polysaccharides, 1.3% total starch, 43% total sugar, 3.8% protein and 4.5% total phlotorannin	Increase of total bacteria, *Bifidobacterium*, *Lactobacillus*	Increase in total SCFA, acetic and propionic acids	[85]

Table 1. *Cont.*

Type of Study	Seaweed, Dosage and Time of Exposure	Polysaccharides Characterization	Significant Changes in Gut Microbiota	Significant Changes in Related Metabolites	Reference
In vitro fermentation system using fresh fecal samples from three3 healthy donors	Polysaccharides extracts obtained by microwave-intensified enzymatic process from 4.5 g of crude *E. radiata* for 24 h. Four different seaweed fractions were employed (crude extract fraction, phlorotannin-enriched fraction, low molecular weight polysaccharide-enriched fraction and high molecular weight polysaccharide-enriched fraction. Inulin was used as positive control and cellulose as negative control	Crude extract fraction: 14.4%, fibre, 5.6% non-digestible non-starch polysaccharides, 20.6% sugar, 0.2% ManA, 0.5% Man, 17.2% Glc, 0.5% Gal, 0.3% Xyl, 1.8% Fuc, 4.6% phlorotannin. Phlorotannin-enriched fraction: 3.4% fibre, 3.4% sugar, 3.4% Glc, 13.4% phlorotannin. Low molecular weight polysaccharide-enriched fraction: 0.5% fibre, 0.4% starch, 22.7% sugar, 22.7% Glc, 2.5% phlorotannin. High molecular weight polysaccharide-enriched fraction: 62.4% fibre, 22.8% non-digestible non-starch polysaccharides, 0.3% starch, 42.1% sugar, 1.9% GulA, 7.2% ManA, 2.1% Man, 1.1% GlcA, 17.1% Glc, 1.7% Gal, 1.5% Xyl, 9.4% Fuc, 1.7% phlorotannin.	Increase of *Bifidobacterium*, *Lactobacillus*, *Clostridium coccoides* in all tested fractions with respect to negative controls. Low molecular weight polysaccharide-enriched fraction showed the better fermentative results, obtaining better counts that positive controls for *Lactobacillus*, *Faecalibacterium prausnitzii*, *C. coccoides* and Firmicutes	Total SCFA were higher in crude fraction than all other fractions after 24 h fermentation. All fractions except phlorotannin-enriched fraction significantly increased SCFA production with respect to negative controls	[7]
In vivo trail using 10 C57BL6 mice per group, with previously induced colitis by supplementing 3% *w/v* of dextran sulphate sodium in the drinking water for 8 days	Fucoidans extracted from *Fucus vesiculosus* intraperitoneally (10 mg/kg/day) or orally (10 mg/kg/day for high purity fucoidan or 400 mg/kg/day for focus-polyphenol) for 7 days	Fucus-polyphenol: 40.2% neutral carbohydrates; 21.8% sulfates; 26.2% polyphenols; 3.6% uronic acids and 203.1 kDa peak molecular weight.High purity fucoidan: 59.5% neutral carbohydrates; 26.6% sulphates; <0.5% polyphenols; 1.4% uronic acids; 61.8 kDa molecular weight.	Not provided	Both oral fucoidan reduced cytokines associated with inflammatory bowel disease such as interleukin-1α, interleukin-1β, interleukin-10, macrophrage inflammatory protein-1α, macrophrage inflammatory protein-1β, granulocyte colony-stimulating factor or granulocyte-macrophage colony-stimulating factor	[23]
In vitro fermentation system using fresh fecal samples from three healthy donors	0.8 g of fucoidans obtained from *Laminaria japonica* for 48 h. Blank samples contained no polysaccharide	Not provided	Decrease in *Enterobacter* spp. while increase in beneficial bacteria as *Lactobacillus* and *Bifidobacterium*	Decrease in pH and increase in lactic acid and SCFA, including acetic and butyic acids	[82]

Table 1. *Cont.*

Type of Study	Seaweed, Dosage and Time of Exposure	Polysaccharides Characterization	Significant Changes in Gut Microbiota	Significant Changes in Related Metabolites	Reference
In vivo trial using 18 male C57BL/6 mice. Six mice received fucoidans extracted *Laminaria japonica*, and six acted as blank group	100 mg/kg/day of fucoidans obtained *L. japonica* for 6 weeks. Control group received saline solution	Fucoidans from *L. japonica* 18.4% sulfate content; 310 KDa molecular weight; 11.2% Man, 7.3% GlcA, 5.2% Glc, 19.3% Gal, 2.9% Xyl, 54.1% Fuc	Increase in the abundance of *Ruminococcaceae*	Decreased in the serum levels of lipopolysaccharide-binding protein	[95]
In vivo trial using six male Wistar rats per group	2% *w/w* of laminarins for 2 weeks. Black samples received control diet	Not provided	Increase of *Bacteroides capillosus*, *Clostridium ramosum* y *Parabacteroides distasonis*	Increase organic acids, specially propionate, whereas decreased cecal putrefactive compounds (indole, phenol and H_2S)	[101]
In vitro fermentation system using fresh fecal samples from three healthy donors and an in vivo trial using 20 Wistar rats	1 g laminarins from *Laminaria digitata* for 24 h. Glucose was used as negative control and FOS as positive control	Not provided	No significant differences were obtained in the in vitro trial for GM composition.	Increase in total SCFA in laminarin-added culture medium than in glucose-added. Laminarins supplementation increased the colon luminal content of mucin, while decreased luminal mucin in jejunum, ileum and caecum in rats	[89]
In vivo trial using 28 female Sprague-Dawley rats	Supplementation with 10% of dried *L. japonica* for 4 weeks. Control rats were fed with basal diet	Not provided	Reduction in Firmicutes to Bacteroidetes ratio and decrease of pathogenic bacteria such as *Clostridium*, *Escherichia* and *Enterobacter*	Increase in total SCFA, and butyric acid. Lower production of acetic acid propionic acids	[90]
In vivo trial using six female BALB/C mice per group	Mice received normal diet, high-fat diet or high-fat diet added with laminarins at 1% *w/w* in a high-fat diet ad libitum for 4 weeks. After finishing, highly-fat diet was provided for an additional 2 weeks	Not provided	Decrease in Firmicutes and increase in Bacteroidetes phylum, especially the genus *Bacteroides* in laminarin-added fed mice with respect to controls	Mice fed with laminarin supplementation showed significantly higher CAZyme families in feces	[44]
In vitro fermentation system using fresh fecal samples from three healthy donors	Polysaccharides isolated from *Laminaria digitata* crude or depolymerized (1% *w/v* for 48 h). Cellulose was used as negative control and FOS as positive control	Not provided	Increase *Parabacteroides*, *Fibrobacter* and *Lachnospiraceae* and decrease in *Streptococcus*, *Ruminococcus* and Peptostreptococcaceae in laminarin-added samples	Increase in SCFA with respect to cellulose-added samples, but similar SCFA content or even lower with respect to FOS-added samples	[91]

Table 1. *Cont.*

Type of Study	Seaweed, Dosage and Time of Exposure	Polysaccharides Characterization	Significant Changes in Gut Microbiota	Significant Changes in Related Metabolites	Reference
In vitro fermentation system with individual bifidobacteria including *B. infantis* JCM 1222; *B. longum* JCM 1217 and *B. adolescentis* JCM 1275	Beta-glucans from *L. digitata* (0.5% *w/v* for 24 h) compared to barley β-glucan, Curdlan from *Alcaligenes faecalis*, mushroom sclerotia from *Pleurothus tuber-regium* and inulin	β-Glucan > 95%, protein 3%; monosaccharides: 98% Gluc; 2% Man. 6 kDa as average molecular weight	Increase of all *Bifidobacteria* with respect to initial counts in a similar way of the other beta-glucans assayed	Increase of SCFA, acetic propionic and butyric acids and decrease of pH in a similar way of the other beta-glucans assayed	[92]
In vitro fermentation system in cellular lines using human-enterocyte-like-29-Luc cells	Supplementation with 0.5% *w/v* and 0.1% *w/v* of sodium alginate and laminarins extracted from *Eisenia bicyclis* for 18 h.	Glu residues with degree of polymerization between 22 and 25 and 5 kDa as average molecular weight	Inhibition of *Salmonella* Typhimurium, *Listeria monocytogenes* or *Vibrio parahaemolyticus* adhesion and invasion	Not provided	[73]
In vivo trial using 24 male C57BL/6J mice	Polysaccharides extracted from *Porphyra haitanensis* (250 mg/kg) for 2 weeks. Control mice received 0.9% normal saline at a dose of 20 mL/kg/day. Positive controls received the same plus combined *Bifidobacterium*, *Lactobacillus* and *Streptococcus thermophilus* tables, 500 mg/kg.	Not provided	Increase of *Prevotellaceae Rikenellaceae* and *Lactobacillus*, while decreased *Lachnoclostridium* or *Lachnospiraceae*	Not provided	[97]
In vivo trial using 16 male C57BL/6 mice fed with a high-fat diet	5% *w/w* polysaccharides extracted from *Saccorhiza polyschides* with high-fat diet for 8 months	Not provided	Not provided	Reduced activity of fecal bile salt hydrolase activity and secondary bile acids	[96]
In vivo trial using Syrian golden hamsters	150 mg/kg body weight of *Sargassum confussum* solution once daily for 60 days by intragastric administration	Sulfated oligosaccharide containing galactose, sulfated galactose, sulfated anhydrogalactose and methyl sulfated galactoside	Increased gut bacterial diversity in treated hamsters. Significant increase in *Barnesiella*, *Tannerella*, *Eubacterium* and *Clostridium* XIVa, with significant decrease in *Allobaculum*, *Bacteroides*, and *Clostridium* IV in the *S. confussum*-added group.	*S. confussum* administration significantly reduced the gene expression of JNK1 and JNK2 in hepatic cells and increased expression of IRS1 and PI3K	[15]
In vitro fermentation system using fresh fecal samples from three healthy donors	Extracts from *Sargassum multicum* (1% *w/v* for 24 h). FOS was used as positive control and no carbon source was added to the negative control	Not provided	Increase *Bacteroides* and *Prevotella*, and decrease in *Clostridium coccoides* and *Eubacterium rectale*	Increase in SCFA and lactic acid production with respect to negative controls	[88]
In vitro test comparing the growth of *L. plantarum* NCIM 2083 with respect to *Salmonella* Typhimurium MTCC 3224	1% *w/v* of enzymatic-extracted polysaccharides from *Sargassum wightii* in MRS broth for 48 h	53.5% fiber, 13.2% protein, 2.3% fat, 28.9% ash. Content of Cel, Fru and Gluc (not specific proportions)	Prebiotic effect by 1.42-fold more growth stimulation of *L. plantarum* than *Salmonella* Typhimurium	Not provided	[86]

Table 1. *Cont.*

Type of Study	Seaweed, Dosage and Time of Exposure	Polysaccharides Characterization	Significant Changes in Gut Microbiota	Significant Changes in Related Metabolites	Reference
In vitro fermentation system using fresh fecal samples from three healthy donors	200 mg polysaccharides extracted from *Sargassum thunberguii* for 48 h	68.3% carbohydrate; 0.3% protein; 3.5% sulfate: Monosaccharide molar ratio: 3.9% arabinose; 6.2% Gal, 3.2% Glc, 15.6% Xyl, 14.8% Man 15.6% GulA, 40.6% GlcA. Average molecular weight 4.8 kDa	Decrease of Firmicutes, while increase of Bacteroidetes and beneficial bacteria such as *Bifidobacterium*, *Roseburia, Parasutterella* and *Fusicatenibacter* after 24-h fermentation, and increase of *Faecalibacterium* and *Coprococcus* after 48-h fermentation	Decrease of pH and increase in total SCFA and acetic, propionic, butiric and *n*-valeric acids	[87]
In vivo trial using 28 female Sprague-Dawley rats	Supplementation with 10% of dried *Undaria pinnatifida* and for 4 weeks. Control rats were fed with basal diet	Not provided	Reduction in Firmicutes to Bacteroidetes ratio and decrease of pathogenic bacteria such as *Clostridium, Escherichia* and *Enterobacter*	Increase in total SCFA, and butyric acid. Lower production of acetic acid propionic acids	[90]

BCC: Biotec Culture Collection; BCC: British Culture Collection; Cel: Cellobiose; Gal: galactose; GlcA: galacturonic acid; Glc: glucose; GulA: guluronic acid; FOS: fructooligosaccharides; Fru: fructose; Fuc: fucose; JCM: Japan Collection of Microorganism; kDa: kilodaltons; ManA: Mannuronic acid; Man: mannose; MRS: Man, Rogosa and Sharpe; MTCC: Microbial Type Culture Collection and Gene Bank; NCIM: National Centre of Integrative Medicine; SCFA: Short chain fatty acids; Xyl: Xylose.

Evidence that probiotic bacteria in the gastrointestinal tract utilize dietary alginate was reviewed by Shang et al. [38]. Among the studies, Kuda et al. [73] found that supplementation with sodium alginate and laminarin of brown algae inhibited the adhesion and invasion of pathogens, such as *S. Typhimurium*, *Listeria monocytogenes* or *Vibrio parahaemolyticus*. Other authors reported an increase in *Lactobacillus* and *Ruminococcus* in the intestine of mice fed fucoidans from *A. nodosum*, besides a reduction in the opportunistic *Peptococcus* bacteria [95].

In vitro fermentation experiments conducted by Charoensiddhi et al. [85] demonstrated the growth-promoting effect of *E. radiata* extracts on beneficial bacteria, such as *Bifidobacterium*, *Lactobacillus* and *Clostridium coccoides*, and SCFAs production was stimulated as well. Later, the same authors [7] found increased levels of beneficial bacteria, such as *Bifidobacterium*, *Lactobacillus* and *C. coccoides* associated with the phlorotannin-enriched fermentation of *E. radiata*. Higher numbers of *Lactobacillus*, *Faecalibacterium prausnitzii*, *C. coccoides*, Firmicutes and *E. coli* were observed for phlorotannin-supplemented fermentation compared with inulin fermentation [7]. In contrast, the number of *Enterococcus* in both fermentations decreased approximately ten-fold relative to the initial counts.

Other authors tested the effects of supplementation of two brown algae (*U. pinnatifida* and *L. japonica*) on the GM and body status of laboratory rats [90]. In both instances, the animals' body weight was reduced, which was thought to be mediated by the influence of the seaweed on the composition of the intestinal microbial communities associated with obesity, reducing the proportion of Firmicutes with respect to Bacteroidetes, and the populations of pathogenic bacteria, such as *Clostridium*, *Escherichia* and *Enterobacter* [90]. Similarly, *L. japonica* increased beneficial bacteria and SCFA, and decreased the pH level [82], while β-glucans extracted from *L. digitata* increased *Bifidobacterium* and propionic and butyric acids in vitro, in addition to lowering pH [92].

β-Glucans obtained from another *Laminaria* species (*L. digitata*) were able in an in vitro test [92] to increase *Bifidobacterium* and propionic and butyric acids, in addition to lowering pH. A study by Strain et al. [91] in vitro investigated the effect of a polysaccharide-rich raw extract obtained from *L. digitata*. A significant alteration of the relative abundance of several families, including *Lachnospiraceae* and genera such as *Streptococcus*, *Ruminococcus* and *Parabacteroides* of human faecal bacterial populations was seen. Concentrations of acetic acid, propionic acid, butyric acid and total SCFA were significantly higher.

Finally, Huebbe et al. [96] conducted a study on mice that were administered polysaccharides from *S. polyschides* with a high-fat diet. A metabolic improvement was seen including normalization of blood glucose, reduction of plasma leptin, reduction of fecal bile salt hydrolase activity and secondary bile acids in these mice.

4.2. Polysaccharides from Red Seaweeds

The most relevant results obtained from the investigation of red seaweeds effect on GM can be found in Table 2. In this table, results about the prebiotic effect of red seaweed species from genus *Acanthopora* [86], *Gracilaria* [105,106], *Kappaphycus* [107], *Euchema* and *Grateloupia* [10], *Chondrus* [57], *Gelidium* [94], or *Osmundea* [88] were included.

As can be seem in Table 2, as was described previously for brown seaweeds, administration of red seaweeds or seaweed-extracted polysaccharides resulted in an increase of SCFA production, stimulating of beneficial bacteria grown such as *Lactobacillus* [86] or *Bifidobacterium* [57,94,107], whereas inhibited the growth of potentially pathogen bacteria [57,86]. It was also reported red seaweeds activity on the prevention of naproxen-induced gastrointestinal damage [106].

Agarose stands out among the polysaccharides isolated from red algae that cannot be digested by human intestinal enzymes. When seaweed is consumed, whether as an edible food or food additive, agarose reaches the most distal portions of the gastrointestinal tract, where it is fermented and metabolized by the GM [108,109]. As described by Ramnani et al. [94], low-molecular-weight agarose exerted a prebiotic effect in vitro by promoting the growth of *Bifidobacterium* and increasing SCFA concentrations in the medium.

Table 2. Prebiotic effect of different species of red seaweed.

Type of Study	Seaweed and Dosage	Polysaccharides Characterization	Significant Changes in Gut Microbiota	Significant Changes in Metabolites	Reference
In vitro test comparing the growth of *Lactobacillus plantarum* NCIM 2083 with respect to *Salmonella* Typhimurium MTCC 3224	1% *w/v* of enzymatic-extracted polysaccharides from *Acanthopora spicifera* in MRS broth for 48 h	45.9% fiber, 10.9% protein, 1.6% fat, 39.4% ash. Only Glu was found as monosaccharide	Prebiotic effect by 0.84-fold more growth stimulation of *L. plantarum* than *S. Typhimurium*	Not provided	[86]
In vivo trial using male Sprague-Dawley rats (six per group)	Fed added with 0.5–2.5% (*w/w*) whole *Chondrus crispus* for 21 days	Not provided	Increase of *Bifidobacterium brevis* and decrease of pathogens such as *Clostridium septicum* and *Streptococcus pneumoniae*	Increase in total SCFA and acetic, propionic and butiric acids in rats fed with *C. crispus* at 0.5% and 2.5%. Higher concentrations of all SCFA were found in the case of rats fed added 2.5% of *C. crispus* with respect to rats fed added with 0.5% of *C. crispus*	[57]
In vitro test in MRS broth for *Lactobacillus* and *Bifidofacterium* compared to MHB broth for *Staphylococcus aureus* and *Escherichia coli*	0.1–0.5% *w/v Eucheuma spinosum* for 24 h	62.1% sugar; 21.4% sulfate; Monosaccharides at molar ratio: 0.01 Man, 0.01 GluA, 1% Gal, 0.09% Xyl, 0.01% Fuc, 0.03% Glu	Increase in beneficial bacteria with better results at 0.1% concentration. No inhibition was detected against pathogens	Not provided	[10]
In vitro test in MRS broth for *Lactobacillus* and *Bifidofacterium* compared to MHB broth for *S. aureus* and *E. coli*	0.1–0.4% *w/v Grateloupia filicina* added in culture media for 24 h	41.9% sugar; 20.6% sulfate; Monosaccharides at molar ratio: 0.01 Man, 0.02 GluA, 1% Gal, 0.1% Xyl, 0.05% Fuc, 0.07% Glu	Increase in beneficial bacteria at all concentrations, without significant differences between 0.4% and 0.5%. No inhibition was detected against pathogens	Not provided	[10]
In vivo trial using male Wistar rats (six per group)	Rats were pretreated with 0.5% carboxymethylcelloluse (controls) or 0.5% *w/v* of of sulphated polysaccharides from *Gracilaria birdiae*, twice daily for 2 days. After 1 h, naproxen (80 mg/kg) was administered twice a day for 2 days	Molar mass distribution was found to be within 2.6×10^6 and 3.8×10^5 g/mol, while the soluble carbohydrate, protein, and sulfate contents were 85.5%, 2.5%, and 8.4%, respectively	No relevant variation was observed in GM populations	Prevention of naproxen-induced gastrointestinal damage determined by macro- and microscopic findings	[106]
In vitro fermentation system using fresh fecal samples from four healthy donors	100 mg of sulphated polysaccharides obtained from *Gracilaria rubra* for 24 h. Basal nutrient medium was used for control negative group and FOS was used for control positive group	Average molecular weight 923.3 kDa, sugar content 0.11%	Increase of Bacteroidetes, *Bacteroidaceae*, *Prevotellaceae*, *Ruminococcaceae* and propionic acid, while decrease *Fusobacteriaceae* and *Lachnospiraceae*. At genus level, increase of *Bacteroides*, *Prevotella* and *Phascolarctobacterium*	Increase in total SCFA and acetic, propionic and isobutyric acids	[105]

Table 2. *Cont.*

Type of Study	Seaweed and Dosage	Polysaccharides Characterization	Significant Changes in Gut Microbiota	Significant Changes in Metabolites	Reference
In vitro fermentation system using fresh fecal samples from three healthy donors	1% *w/v* low molecular weight polysaccharide derivatives extracted from *Gracilaria* spp. for 24 h. Inulin was used as positive control and cellulose as negative control	Average molecular weight 143.8 kDa	No significant changes in GM	Increase in total SCFA, acetic and propionic acids	[94]
In vitro fermentation system using fresh fecal samples from three healthy donors	1% *w/v* low molecular weight polysaccharide derivatives extracted from *Gelidium sesquipidale* for 24 h. Inulin was used as positive control and cellulose as negative control	Average molecular weight 20.1 kDa and 6.5 kDa, respectively	Only *G. sesquipidale* of 6.5 kDa significantly increased *Bifidobaterium* counts	Both *G. sesquipidale* extracts (20.1 kDa and 6.5 kDa molecular weight) significantly increased total SCFA, acetic and propionic acids	[94]
In vitro fermentation system	1% *w/v Kappaphycus alvarezii* for 24 h	Not provided	Increase in *Bifidobacterium*, decrease in *Clostridium coccoides* and *Eubacterium rectale*	Increase in SCFA	[107]
In vitro fermentation system using fresh fecal samples from threehealthy donors	Extracts from *Osmundea pinnatifida* (1% *w/v* for 24 h). FOS was used as positive control and no carbon source was added to the negative control	Not provided	Increase in *Bifidobaterium* counts	Increase in SCFA, acetic and propionic acids	[88]

Gal: galactose; Glc: glucose; GulA: guluronic acid; FOS: fructooligosaccharides; Fuc: fucose; GM. Gut microbiota; kDa: kilodaltons; Man: mannose; MRS: Man, Rogosa and Sharpe; MTCC: Microbial Type Culture Collection and Gene Bank; NCIM: National Centre of Integrative Medicine; SCFA: Short chain fatty acids; Xyl: Xylose,

Bajury et al. [107] conducted an in vitro colon model in which they evaluated the prebiotic capacity of *K. alvarezii*. This study showed an increase in SCFA (particularly acetate and propionate) and *Bifidobacterium*. In the other hand, decrease in *C. coccoides* and *E. rectale*. These results suggested that *K. alvarezii* might have the potential as a prebiotic ingredient. A study published by Zhang et al. [110] focused on the beneficial effect of low-melting-point agarose (in the form of neoagaro-oligosaccharides) on the GM during the relief of intense exercise-induced fatigue in mice. Results showed the abundance of Bacteroidetes and Proteobacteria increased and decreased, respectively, during the attenuation of fatigue and its associated gastrointestinal problems. Ladirat et al. [111] found that mice fed 2.5% (*w/v*) neoagaro-oligosaccharides for seven consecutive days achieved a much more pronounced increase in the population of *Lactobacillus* spp. and *Bifidobacterium* spp. in their GM relative to those fed 5% (*w/v*) FOS for 14 consecutive days. Likewise, it was demonstrated that agaro-oligosaccharides could be used as a prebiotic to encourage the growth of beneficial strains of bacteria, such as *B. adolescentis* ATCC 15703 and *B. infantis* ATCC 15697. Low-molecular-weight agar has demonstrated a bifidogenic effect, along with an increase in SCFA acetate and propionate concentrations, after 24 h of in vitro fermentation with human faeces inoculant [94].

Another type of polysaccharide with prebiotic function found in red algae are the group of carrageenans, which are derived from D-galactose, and approved as food additives [79]. In rats fed 2.5% *C. crispus*, of which carrageenan is a major polysaccharide, *B. brevis*, as well as SCFA, increased considerably, while pathogens *Clostridium septicum* and *Streptococcus pneumonia* noticeably decreased compared with the basal diet [57]. Elevation of plasma immunoglobulin levels was also found in rats fed with *C. crispus*, resulting in improved host immunity. Consistent with the prebiotic activity of carrageenan, carrageenans isolated from red algae *G. filicina* and *E- spinosum* promoted the growth of *Bifidobacterium* [10].

Research led by Di et al. [105] found that the polysaccharides of *Gracilaria rubra* increased the relative abundances of *Bacteroides*, *Prevotella* and *Phascolarctobacterium* in vitro compared with the control group. *Bacteroides* spp. assists the host with degrading polysaccharides and contains codifying genes of glucosidase enzymes [39]. *Prevotella* is another beneficial genus with the potential to participate in the metabolism and utilization of plant polysaccharides. The genus *Phascolarctobacterium* is associated with the production of SCFA [110].

Many other bacterial genera, such as *Legionella*, *Sutterella*, *Blautia*, *Holdemania*, *Shewanella* and *Agarivorans*, were decreased as a consequence of intake of *C. crispus* supplements in rats [57]. Decreases in the presence of *Streptococcus* were also observed. In conclusion, carrageenans from *C. crispus* could act as a fermentable substrate for probiotic bacteria present in the gastrointestinal tract, thereby promoting the growth of probiotic groups, while inhibiting certain groups of pathogenic bacteria [57]. Another study in chickens described an overall impact of administering whole red algae (*Sarcodiotheca gaudichaudii* and *C. crispus*) on the intestinal mucosa, increasing the height and surface of the villi in these animals [112]. Moreover, the abundance of beneficial bacteria, such as *B. longum* and *Streptococcus salivarius* increased, while some harmful bacteria species, such as *C. perfringens*, decreased [112] Rodrigues et al. [88] used extracts from the red algae *O. pinnatifida* and *S. muticum* in an in vitro fermentation system, which increased the production of acetate and propionate, and the population of *Bifidobacterium*. In work published by Silva et al. [106], in which extracts of sulphated polysaccharides from *G. birdiae* were administered to laboratory rats, gastrointestinal damage induced by naproxen was prevented, although it did not produce notable variations in the GM of these rats.

An in vitro study using *A. spicifera* in MRS broth evaluated their antioxidant activity and prebiotic score with *L. plantarum* and *S. Typhimurium*. The study showed that the prebiotic activity score was positive, promoting the growth of *L. plantarum* and suppressing the growth of the pathogen *S. Typhimurium*. Specifically, a prebiotic effect by 0.84-fold more growth stimulation of *L. plantarum* than *S. Typhymurium* [86].

4.3. Polysaccharides from Green Seaweeds

Unlike brown and red algae, the current evidence for the fermentation capacity of green algae and their polysaccharides is scarce, partly because their fermentation requires a specific activity of α-ʟ-rhamnosidase in the gastrointestinal tract, which is infrequent [113]. The most relevant results obtained from the investigation of green seaweeds effect on GM can be found in Table 3. In this table, results about the prebiotic effect of green seaweed species from genus *Enteromorpha* [38,82,86,113–115] and *Ulva* [115] were included. Administration of green seaweeds or seaweed-extracted polysaccharides also resulted in an increase of SCFA production, stimulating of beneficial bacteria grown such as *Lactobacillus* [38,86,115], *Bifidobacterium* [38], or *Akkermansia* [38] whereas inhibited the growth of potentially pathogen bacteria [81,114]. Other beneficial actions were reported such as decrease lipopolysaccharide-binding protein in female mice [38], diminished histopathological lesions of inflammatory infiltrations in distal colon [114], or modulating diabetes-related genes expression in diabetic mice [22].

Ulvans are one of the most frequent polysaccharides in green algae. This polysaccharide is a water-soluble sulphated heteropolysaccharide [79]. Ulvans contains sulphate and uronic acids, and so produce undigestible ionic colloids, has ion-exchange capacity and can bind to bile acids, consequently increasing the excretion of bile acids with cholesterol-lowering or antihyperlipidemic effects [2,79]. Antioxidant and immunomodulatory properties are other beneficial actions elicited by ulvans [116,117].

Ulvans has also been studied for its possible prebiotic potential. Kong et al. [82] performed an in vitro assay using *Enteromorpha* with a high content of ulvans, but there were no noticeable variations in the populations of *Enterococcus*, *Lactobacillus* and *Bifidobacterium* compared with controls. In contrast, in a recent in vitro faecal fermentation analysis, ulvans stimulated the growth of *Bifidobacterium* and *Lactobacillus* populations and promoted the production of SCFA, such as lactic and acetic acids [42]. In a previous study by Ren et al. [114], both whole *Enteromorpha* and polysaccharides extracted from *Enteromorpha* improved inflammation associated with loperamide-induced constipation in mice. In those mice, the GM showed an increase in Firmicutes and Actinobacteria compared with the control mice, whereas the relative amounts of Bacteroidetes and Proteobacteria decreased.

In work by Shang et al. [38], an extract of *E. clathrata* was administered to mice, resulting in marked decreased concentrations of genera, such as *Enterobacter*, *Staphylococcus* and *Streptococcus*. Surprisingly, such supplementation also dramatically reduced the population of *A. muciniphila* in the intestine. These observations indicate a possible unfavorable effect of these polysaccharides on the GM. Contrarily, these polysaccharides were reported to increase the abundance of *A. muciniphila*, *Bacteroides*, *Alloprevotella*, *Ruminococcaceae* and *Blautia* in the intestinal tract of mice, and decrease the abundance of *Peptococcus*, *Rikenellaceae* and *Alistipes* [96,109].

An in vitro faecal fermentation of xylans derived from *Palmaria palmata* reported that xylose was fermented after 6 h, and the SCFA content increased simultaneously [32]. This study did not determine the bacterial composition. Nonetheless, Xylans and xylo-oligosaccharides extracted from terrestrial plants, such as wheat husks and corn, are considered potential prebiotics due to evidence of bifidogenesis, improved plasma lipid profile and positive modulation of immune function markers in healthy adults [118].

Table 3. Prebiotic effect of different species of green seaweed.

Type of Study	Seaweed, Dosage and Time of Exposure	Polysaccharides Characterization	Significant Changes in Gut Microbiota	Significant Changes in Metabolites	Reference
In vivo trial using 36 C57BL/6J mice, 18 males and 18 females in different trials	*Enteromorpha clathrata*, 100 mg/kg/day or 50 mg mg/kg/day for 4 weeks	Molecular weight 11.67 kDa; 14.7% sulfate content. Monosaccharide composition: 1.0% Man, 49.7% Rha, 10.8% GlcA; 29.9% Glc; 1.3% Gal; 7.2% Xyl	Increase of *Akkermansia muciniphila*, *Bifidobacterium* spp., and *Lactobacillus* spp. *E. clathrata* supplementation induced much less alteration in the composition of female GM than in male GM	*E. clathrata* supplementation decreased lipopolysaccharide-binding protein in female mice but not in male mice	[38]
In vitro test comparing the growth of *Lactobacillus plantarum* NCIM 2083 with respect to *Salmonella* Typhimurium MTCC 3224	1% *w/v* of enzymatic-extracted polysaccharides from *Enteromorpha compressa* in MRS broth for 48 h	60.6% fiber, 16.9% protein, 1.2% fat, 25.4% ash. Monosaccharide content included cellobiose, fructose, glucose and maltose	Prebiotic effect by 1.44-fold more growth stimulation of *Lactobacillus plantarum* than *Salmonella* Typhimurium	Not provided	[86]
In vivo trial using 24 male C57BL/6J mice	Polysaccharides extracted from *Ulva prolifera* (250 mg/kg) for 2 weeks. Control mice received 0.9% normal saline at a dose of 20 mL/kg/day. Positive controls received the same plus combined *Bifidobacterium*, *Lactobacillus* and *Streptococcus thermophilus* tables, 500 mg/kg.	Not provided	Polysaccharides supplementation decreased Tenericutes, and Cyanobacteria. At genus level, decreased *Lachnospiraceae*, *Lactobacillus*, *Mollicutes* and *Mucispirillum*, while increased *Prevotellaceae* and *Rickenellaceae*	Not provided	[115]
In vitro fermentation system using fresh fecal samples from three healthy donors	0.8 g of fucoidans obtained from *Enteromorpha prolifera* for 48 h. Blank samples contained no polysaccharide	Not provided	Decrease in *Enterobacter* spp. in *E. prolifera* fucoidans-added samples	Not significant changes	[82]
In vivo trial using 24 Kunming female mice	Loperamide at a dosage of 9.6 mg/kg/twice a day via oral gavage for 2 weeks was provided to mice to induce slow-transit constipation in mice. Afterwards, *E. prolifera* and polysaccharides extracted from *E. prolifera* added in fed at a 1:5 *w/v* ratio was administered for 7 days	Not provided	*E. prolifera* increased bacterial diversity, Bacteroidales, Firmicutes, Actinobacteria, and decreased Bacteroidetes and Proteobacteria. Extracts from *E. prolifera* increased *Prevotellaceae*, Firmicutes, Actinobacteria, and decreased Bacteroidetes and Proteobacteria	Both *E. prolifera* and *E. prolifera* extracts diminished histopathological lesions of inflammatory infiltrations in distal colon. Both *E. prolifera* and *E. prolifera* extracts reduced serum levels of nitric oxide (inhibitory neurotransmitter) and showed laxative effects	[114]
In vivo trial using 24 Kunming male mice	Treated mice were fed with high sucrose/high fat diet for 5 weeks. Next type-2 diabetes was induced by intraperitoneal administration of streptozotocin at 45 mg/kg for 3 days. Diabetic mice were administered with 150 mg/kg *E. prolifera* extracts or its flavonoid-rich fractions less than 3 kDa, respectively, for 4 weeks	Not provided	*E. prolifera* extracts increased the proportion of *Alistipes*, *Lachnospiraceae* and *Odoribacter*, while both extracts reduced the proportion of *Ruminiclostridium* and *Akkermansia* in GM of diabetic mice	Flavonoids from *E. prolifera* reduced blood glucose in mice, reduced mRNA expressions of JNK1/2 gene and increased the expression of PI3K, IRS1 and AKT genes in diabetic mice	[22]

Gal: galactose; GlcA: glucuronic acid; Glc: glucose; GM: gut microbiota; kDa: kilodaltons; Man: mannose; MRS: man Rogosa and Sharpe; MTCC: Microbial Type Culture Collection and Gene Bank; NCIM: National Centre of Integrative Medicine; Rha: rhamnose; SCFA: Short chain fatty acids; Xyl: Xylose.

E. clathrata is an edible green seaweed possessing polysaccharides with numerous bioactivities, including anticoagulant, immunomodulatory, antioxidant, anticancer and anti-obesity effects [38]. It was reported that the polysaccharides of *E. clathrata* exerted diverse prebiotic effects on *A. muciniphila*, *Bifidobacterium* and *Lactobacillus* in male and female mice [38]. The results were most evident in the male mice because of a sex-specific effect on the GM, as sex hormones play a key role in determining the composition of intestinal microorganisms [109]. In other work from the same authors, male mice were supplemented with polysaccharides of *E. clathrata* in the diet, which increased the abundance of *Bacteroides, Prevotella, Alloprevotella, Eubacterium* and *Peptococcus*, and decreased the proportion of the cancer-related *Helicobacter* [109]. In the female counterparts, the abundance of *Odoribacter, Clostridium* IV, *Oscillibacter* and *Alistipes* spp. increased, and the proportions of beta-proteobacteria decreased [38].

An in vitro study using *E. compressa* in MRS broth evaluated their antioxidant activity and prebiotic score with *L. plantarum* and *S. Typhimurium*. The study showed that the prebiotic activity score was positive, promoting the growth of *L. plantarum* and suppressing the growth of the pathogen *S. Typhimurium*. This seaweed exhibited the highest score of prebiotic activity (1.44-fold), stimulating the growth of *L. plantarum* than *S. typhimurium* [86].

The natural products of marine macroalgae have shown notable antidiabetic potential by interfering with carbohydrate metabolism. For example, *E. prolifera* contains many bioactive compounds, such as sulphated polysaccharides, which could improve glucose metabolism, in addition to displaying anti-inflammatory, antiviral and anticoagulant functions [22].

5. Conclusions

Although substantial evidence of the prebiotic effect of seaweed and seaweed extracts has been published in recent years, these studies have been performed using in vitro digestion systems simulating the human colon, or in animal models. Animals, such as mice or rats, differ widely from humans in the GM composition, immune function, diets, metabolism and other key aspects, so extrapolating the results obtained from animal models to humans may not be valid. In vitro systems replicate more similarly the human intestinal microbiota, but are less-dynamic systems than the real human colonic environment. Additionally, other factors should be considered, such as the possible effect of other secondary compounds contained in seaweeds on the GM composition, or the potential to transfer genes from marine bacteria to human GM bacteria coding for enzymes that could degrade seaweed polysaccharides. Thus, not all people will respond equally after seaweed ingestion. The decrease in terrestrial agriculture and disposable water is likely to increase the consumption of algae by humans in the near future. Meanwhile, there is a profound need for more in-depth investigations into the potential prebiotic effects of marine seaweeds and their derived polysaccharides on the human GM.

Author Contributions: Conceptualization, J.M.M. and A.C. Literature data collection, A.C.A. and A.L.-S. Writing—original draft, A.L.-S. and A.d.C.-M. Writing—review and editing, A.C.-C., C.M.F., and A.L. Supervision: J.M.M. and A.C. All authors have read and agreed to the published version of the manuscript.

Funding: The authors thank the European Regional Development Funds (FEDER), grant ED431C 2018/05, and Programa Iberoamericano de Ciencia y Tecnología para el Desarrollo (CyTED), grant PCI2018-093245 for covering the cost of publication.

Conflicts of Interest: The authors declare no conflict of interest.

References

1. Cian, R.E.; Drago, S.R.; De Medina, F.S.; Martínez-Augustin, O. Proteins and carbohydrates from red seaweeds: Evidence for beneficial effects on gut function and microbiota. *Mar. Drugs* **2015**, *13*, 5358–5383. [CrossRef] [PubMed]

2. Gurpilhares, D.B.; Cinelli, L.P.; Simas, N.K.; Pessoa, A., Jr.; Sette, L.D. Marine prebiotics: Polysaccharides and oligosaccharides obtained by using microbial enzymes. *Food Chem.* **2019**, *280*, 175–186. [CrossRef]

3.   Brown, E.M.; Allsopp, P.J.; Magee, P.J.; Gill, C.I.; Nitecki, S.; Strain, C.R.; McSorley, E.M. Seaweed and human health. *Nutr. Rev.* **2014**, *72*, 205–216. [CrossRef] [PubMed]

4.   Barba, F.J. Microalgae and seaweeds for food applications: Challenges and perspectives. *Food Res. Int.* **2017**, *99*, 969–970. [CrossRef] [PubMed]

5.   Tiwari, B.K.; Troy, D.J. Seaweed sustainability–food and nonfood applications. In *Seaweed Sustainability*; Elsevier: Oxford, UK, 2015; pp. 1–6.

6.   FAO. *The Future of Food and Agriculture–Trends and Challenges*; Food and Agriculture Organization of the United Nations: Rome, Italy, 2017.

7.   Charoensiddhi, S.; Conlon, M.A.; Vuaran, M.S.; Franco, C.M.; Zhang, W. Polysaccharide and phlorotannin-enriched extracts of the brown seaweed *Ecklonia radiata* influence human gut microbiota and fermentation in vitro. *J. Appl. Phycol.* **2017**, *29*, 2407–2416. [CrossRef]

8.   Charoensiddhi, S.; Conlon, M.A.; Franco, C.M.; Zhang, W. The development of seaweed-derived bioactive compounds for use as prebiotics and nutraceuticals using enzyme technologies. *Trends Food Sci. Technol.* **2017**, *70*, 20–33. [CrossRef]

9.   Makkar, H.P.; Tran, G.; Heuzé, V.; Giger-Reverdin, S.; Lessire, M.; Lebas, F.; Ankers, P. Seaweeds for livestock diets: A review. *Anim. Feed Sci. Technol.* **2016**, *212*, 1–17. [CrossRef]

10.   Chen, X.; Sun, Y.; Hu, L.; Liu, S.; Yu, H.; Li, R.; Wang, X.; Li, P. In vitro prebiotic effects of seaweed polysaccharides. *J. Oceanol. Limnol.* **2018**, *36*, 926–932. [CrossRef]

11.   Cardoso, S.M.; Pereira, O.R.; Seca, A.M.L.; Pinto, D.C.G.A.; Silva, A.M.S. Seaweeds as preventive agents for cardiovascular diseases: From nutrients to functional foods. *Mar. Drugs* **2015**, *13*, 6838–6865. [CrossRef]

12.   Kumar, S.A.; Magnusson, M.; Ward, L.C.; Paul, N.A.; Brown, L. Seaweed supplements normalize metabolic, cardiovascular and liver responses in high-carbohydrate, high-fat fed rats. *Mar. Drugs* **2015**, *13*, 788–805. [CrossRef]

13.   Seca, A.M.L.; Pinto, D.C.G.A. Overview of the antihypertensive and anti-obesity effects of secondary metabolites from seaweeds. *Mar. Drugs* **2018**, *16*, 237. [CrossRef] [PubMed]

14.   Wan-Loy, C.; Siew-Moi, P. Marine algae as a potential source for anti-obesity agents. *Mar. Drugs* **2016**, *14*, 222. [CrossRef] [PubMed]

15.   Yang, C.F.; Lai, S.S.; Chen, Y.H.; Liu, D.; Liu, B.; Ai, C.; Wan, W.Z.; Gao, L.Y.; Chen, X.H.; Zhao, C. Anti-diabetic effect of oligosaccharides from seaweed *Sargassum confusum* via JNK-IRS1/PI3K signaling pathways and regulation of gut microbiota. *Food Chem. Toxicol.* **2019**, *131*, 110562. [CrossRef] [PubMed]

16.   Zao, C.; Yang, C.; Liu, B.; Lin, L.; Sarker, S.D.; Nahar, L.; Yu, H.; Cao, H.; Xiao, J. Bioactive compounds from marine macroalgae and their hypoglycemic benefits. *Trends Food Sci. Technol.* **2018**, *72*, 1–12. [CrossRef]

17.   Wang, H.M.D.; Li, X.C.; Lee, D.J.; Chang, J.S. Potential biomedical applications of marine algae. *Bioresour. Technol.* **2017**, *244*, 1407–1415. [CrossRef]

18.   Xue, M.; Ji, X.; Liang, H.; Liu, Y.; Wang, B.; Sun, L.; Li, W. The effect of fucoidan on intestinal flora and intestinal barrier function in rats with breast cancer. *Food Funct.* **2018**, *9*, 1214–1223. [CrossRef]

19.   Shin, T.; Ahn, M.; Hyun, J.W.; Kim, S.H.; Moon, C. Antioxidant marine algae phlorotannins and radioprotection: A review of experimental evidence. *Acta Histochem.* **2014**, *116*, 669–674. [CrossRef]

20.   Chater, P.I.; Wilcox, M.; Cherry, P.; Herford, A.; Mustar, S.; Wheater, H.; Brownlee, I.; Seal, C.; Pearson, J. Inhibitory activity of extracts of Hebridean brown seaweeds on lipase activity. *J. Appl. Phycol.* **2016**, *28*, 1303–1313. [CrossRef]

21.   You, L.; Gong, Y.; Li, L.; Hu, X.; Brennan, C.; Kulikouskaya, V. Beneficial effects of three brown seaweed polysaccharides on gut microbiota and their structural characteristics: An overview. *Int. J. Food Sci. Technol.* **2019**. [CrossRef]

22.   Yan, X.; Yang, C.; Chen, Y.; Miao, S.; Liu, B.; Zhao, C. Antidiabetic potential of green seaweed *Enteromorpha prolifera* flavonoids regulating insulin signaling pathway and gut microbiota in type 2 diabetic mice. *J. Food Sci.* **2019**, *84*, 165–173. [CrossRef]

23.   Lean, Q.Y.; Eri, R.D.; Fitton, J.H.; Patel, R.P.; Gueven, N. Fucoidan extracts ameliorate acute colitis. *PLoS ONE* **2015**, *10*, e0128453. [CrossRef] [PubMed]

24.   Zeid, A.A.; Aboutabl, E.A.; Sleem, A.; El-Rafie, H. Water soluble polysaccharides extracted from *Pterocladia capillacea* and *Dictyopteris membranacea* and their biological activities. *Carbohydr. Polym.* **2014**, *113*, 62–66. [CrossRef] [PubMed]

25. Fedorov, S.N.; Ermakova, S.P.; Zvyagintseva, T.N.; Stonik, V.A. Anticancer and cancer preventive properties of marine polysaccharides: Some results and prospects. *Mar. Drugs* **2013**, *11*, 4876–4901. [CrossRef] [PubMed]

26. Wijesinghe, W.; Jeon, Y. Biological activities and potential industrial applications of fucose rich sulfated polysaccharides and fucoidans isolated from brown seaweeds: A review. *Carbohydr. Polym.* **2012**, *88*, 13–20. [CrossRef]

27. Qi, H.; Huang, L.; Liu, X.; Liu, D.; Zhang, Q.; Liu, S. Antihyperlipidemic activity of high sulfate content derivative of polysaccharide extracted from *Ulva pertusa* (Chlorophyta). *Carbohydr. Polym.* **2012**, *87*, 1637–1640. [CrossRef]

28. Fleita, D.; El-Sayed, M.; Rifaat, D. Evaluation of the antioxidant activity of enzymatically-hydrolyzed sulfated polysaccharides extracted from red algae; *Pterocladia capillacea*. *LWT-Food Sci. Technol.* **2015**, *63*, 1236–1244. [CrossRef]

29. Collins, K.G.; Fitzgerald, G.F.; Stanton, C.; Ross, R.P. Looking beyond the terrestrial: The potential of seaweed derived bioactives to treat non-communicable diseases. *Mar. Drugs* **2016**, *14*, 60. [CrossRef]

30. Torres, M.D.; Flórez-Fernández, N.; Domínguez, H. Integral utilization of red seaweed for bioactive production. *Mar. Drugs* **2019**, *17*, 314. [CrossRef]

31. Wells, M.L.; Potin, P.; Craigie, J.S.; Raven, J.A.; Merchant, S.S.; Helliwell, K.E.; Smith, A.G.; Camire, M.E.; Brawley, S.H. Algae as nutritional and functional food sources: Revisiting our understanding. *J. Appl. Phycol.* **2017**, *29*, 949–982. [CrossRef]

32. Cherry, P.; Yadav, S.; Strain, C.R.; Allsopp, P.J.; McSorley, E.M.; Ross, R.P.; Stanton, C. Prebiotics from seaweeds: An ocean of opportunity? *Mar. Drugs* **2019**, *17*, 327. [CrossRef]

33. O'Sullivan, L.; Murphy, B.; McLoughlin, P.; Duggan, P.; Lawlor, P.G.; Hughes, H.; Gardiner, G.E. Prebiotics from marine macroalgae for human and animal health applications. *Mar. Drugs* **2010**, *8*, 2038–2064. [CrossRef] [PubMed]

34. Jiao, G.; Yu, G.; Zhang, J.; Ewart, H.S. Chemical structures and bioactivities of sulfated polysaccharides from marine algae. *Mar. Drugs* **2011**, *9*, 196–223. [CrossRef] [PubMed]

35. Li, M.; Li, G.; Shang, Q.; Chen, X.; Liu, W.; Zhu, L.; Yin, Y.; Yu, G.; Wang, X. In vitro fermentation of alginate and its derivatives by human gut microbiota. *Anaerobe* **2016**, *39*, 19–25. [CrossRef] [PubMed]

36. Vera, J.; Castro, J.; Gonzalez, A.; Moenne, A. Seaweed polysaccharides and derived oligosaccharides stimulate defense responses and protection against pathogens in plants. *Mar. Drugs* **2011**, *9*, 2514–2525. [CrossRef]

37. Usov, A.I. Polysaccharides of the red algae. In *Advances in Carbohydrate Chemistry and Biochemistry*; Elsevier: San Diego, CA, USA, 2011; Volume 65, pp. 115–217.

38. Shang, Q.; Wang, Y.; Pan, L.; Niu, Q.; Li, C.; Jiang, H.; Cai, C.; Hao, J.; Li, G.; Yu, G. Dietary polysaccharide from *Enteromorpha Clathrata* modulates gut microbiota and promotes the growth of *Akkermansia muciniphila*, *Bifidobacterium* spp. and *Lactobacillus* spp. *Mar. Drugs* **2018**, *16*, 167. [CrossRef]

39. Hehemann, J.; Correc, G.; Barbeyron, T.; Helbert, W.; Czjzek, M.; Michel, G. Transfer of carbohydrate-active enzymes from marine bacteria to Japanese gut microbiota. *Nature* **2010**, *464*, 908. [CrossRef]

40. de Jesus Raposo, M.F.; De Morais, A.M.M.B.; De Morais, R.M.S.C. Emergent sources of prebiotics: Seaweeds and microalgae. *Mar. Drugs* **2016**, *14*, 27. [CrossRef]

41. Fehlbaum, S.; Prudence, K.; Kieboom, J.; Heerikhuisen, M.; van den Broek, T.; Schuren, F.H.J.; Steinert, R.E.; Raederstorff, D. In vitro fermentation of selected prebiotics and their effects on the composition and activity of the adult gut microbiota. *Int. J. Mol. Sci.* **2018**, *19*, 3097. [CrossRef]

42. Seong, H.; Bae, J.; Seo, J.S.; Kim, S.; Kim, T.; Han, N.S. Comparative analysis of prebiotic effects of seaweed polysaccharides laminaran, porphyran, and ulvan using in vitro human fecal fermentation. *J. Funct. Foods* **2019**, *57*, 408–416. [CrossRef]

43. Li, P.; Ying, J.; Chang, Q.; Zhu, W.; Yang, G.; Xu, T.; Yi, H.; Pan, R.; Zhang, E.; Zeng, X.; et al. Effects of phycoerythrin from *Gracilaria lemaneiformis* in proliferation and apoptosis of SW480 cells. *Oncol. Rep.* **2016**, *36*, 3536–3544. [CrossRef]

44. Nguyen, S.G.; Kim, J.; Guevarra, R.B.; Lee, J.; Kim, E.; Kim, S.; Unno, T. Laminarin favorably modulates gut microbiota in mice fed a high-fat diet. *Food Funct.* **2016**, *7*, 4193–4201. [CrossRef] [PubMed]

45. Gomez-Zavaglia, A.; Prieto Lage, M.A.; Jimenez-Lopez, C.; Mejuto, J.C.; Simal-Gandara, J. The potential of seaweeds as a source of functional ingredients of prebiotic and antioxidant value. *Antioxidants* **2019**, *8*, 406. [CrossRef]

46. Okolie, C.L.; CK Rajendran, S.R.; Udenigwe, C.C.; Aryee, A.N.; Mason, B. Prospects of brown seaweed polysaccharides (BSP) as prebiotics and potential immunomodulators. *J. Food Biochem.* **2017**, *41*, e12392. [CrossRef]

47. Freitas, A.C.; Pereira, L.; Rodrigues, D.; Carvalho, A.P.; Panteleitchouk, T.; Gomes, A.M.; Duarte, A.C. Marine functional foods. In *Springer Handbook of Marine Biotechnology*; Springer: Heidelberg, Germany, 2015; pp. 969–994.

48. Rosa, G.P.; Tavares, W.R.; Sousa, P.M.C.; Pagès, A.K.; Seca, A.M.L.; Pinto, D.C.G.A. Seaweed secondary metabolites with beneficial health effects: An overview of successes in in vivo studies and clinical trials. *Mar. Drugs* **2019**, *18*, 8. [CrossRef] [PubMed]

49. Miranda, J.M.; Carrera, M.; Barros-Velázquez, J.; Aubourg, S.P. Impact of previous active dipping in *Fucus spiralis* extract on the quality enhancement of chilled lean fish. *Food Control* **2018**, *90*, 407–414. [CrossRef]

50. Miranda, J.M.; Trigo, M.; Barros-Velázquez, J.; Aubourg, S.P. Effect of an icing medium containing the alga *Fucus spiralis* on the microbiological activity and lipid oxidation in chilled megrim (*Lepidorhombus whiffiagonis*). *Food Control* **2016**, *59*, 290–297. [CrossRef]

51. Miranda, J.M.; Ortiz, J.; Barros-Velázquez, J.; Aubourg, S.P. Quality enhancement of chilled fish by including alga *Bifurcaria bifurcata* extract in the icing medium. *Food Bioprocess Technol.* **2016**, *9*, 387–395. [CrossRef]

52. Oucif, H.; Miranda, J.M.; Mehidi, S.A.; Abi-Ayad, S.E.; Barros-Velázquez, J.; Aubourg, S.P. Effectiveness of a combined ethanol–aqueous extract of alga *Cystoseira compressa* for the quality enhancement of a chilled fatty fish species. *Eur. Food Res. Technol.* **2018**, *244*, 291–299. [CrossRef]

53. Arulkumar, A.; Paramasivam, S.; Miranda, J.M. Combined effect of icing medium and red alga *Gracilaria verrucosa* on shelf life extension of Indian Mackerel (*Rastrelliger kanagurta*). *Food Bioprocess Technol.* **2018**, *11*, 1911–1922. [CrossRef]

54. Nwosu, F.; Morris, J.; Lund, V.A.; Stewart, D.; Ross, H.A.; McDougall, G.J. Anti-proliferative and potential anti-diabetic effects of phenolic-rich extracts from edible marine algae. *Food Chem.* **2011**, *126*, 1006–1012. [CrossRef]

55. Liu, M.; Zhang, W.; Wei, J.; Qiu, L.; Lin, X. Marine bromophenol bis (2,3-dibromo-4,5-dihydroxybenzyl) ether, induces mitochondrial apoptosis in K562 cells and inhibits topoisomerase I in vitro. *Toxicol. Lett.* **2012**, *211*, 126–134. [CrossRef] [PubMed]

56. Liu, F.; Wang, X.; Shi, H.; Wang, Y.; Xue, C.; Tang, Q. Polymannuronic acid ameliorated obesity and inflammation associated with a high-fat and high-sucrose diet by modulating the gut microbiome in a murine model. *Br. J. Nutr.* **2017**, *117*, 1332–1342. [CrossRef] [PubMed]

57. Liu, J.; Kandasamy, S.; Zhang, J.; Kirby, C.W.; Karakach, T.; Hafting, J.; Critchley, A.T.; Evans, F.; Prithiviraj, B. Prebiotic effects of diet supplemented with the cultivated red seaweed *Chondrus crispus* or with fructo-oligo-saccharide on host immunity, colonic microbiota and gut microbial metabolites. *BMC Complement. Altern. Med.* **2015**, *15*, 279. [CrossRef] [PubMed]

58. Chen, L.; Xu, W.; Chen, D.; Chen, G.; Liu, J.; Zeng, X.; Shao, R.; Zhu, H. Digestibility of sulfated polysaccharide from the brown seaweed *Ascophyllum nodosum* and its effect on the human gut microbiota *in vitro*. *Int. J. Biol. Macromol.* **2018**, *112*, 1055–1061. [CrossRef]

59. Kim, S.; Kim, S.R.; Oh, M.; Jung, S.; Kang, S.Y. In vitro antiviral activity of red alga, *Polysiphonia morrowii* extract and its bromophenols against fish pathogenic infectious hematopoietic necrosis virus and infectious pancreatic necrosis virus. *J. Microbiol.* **2011**, *49*, 102–106. [CrossRef]

60. Xu, X.; Piggott, A.M.; Yin, L.; Capon, R.J.; Song, F. Symphyocladins A–G: Bromophenol adducts from a Chinese marine red alga, *Symphyocladia latiuscula*. *Tetrahedron Lett.* **2012**, *53*, 2103–2106. [CrossRef]

61. Abou-El-Wafa, G.; Shaaban, M.; Shaaban, K.; El-Naggar, M.; Maier, A.; Fiebig, H.; Laatsch, H. Pachydictyols B and C: New diterpenes from *Dictyota dichotoma* Hudson. *Mar. Drugs* **2013**, *11*, 3109–3123. [CrossRef]

62. Rubiano-Buitrago, P.; Duque, F.; Puyana, M.; Ramos, F.; Castellanos, L. Bacterial biofilm inhibitor diterpenes from *Dictyota pinnatifida* collected from the Colombian Caribbean. *Phytochem. Let.* **2019**, *30*, 74–80. [CrossRef]

63. Tang, Z.; Gao, H.; Wang, S.; Wen, S.; Qin, S. Hypolipidemic and antioxidant properties of a polysaccharide fraction from *Enteromorpha prolifera*. *Int. J. Biol. Macromol.* **2013**, *58*, 186–189. [CrossRef]

64. Girard, L.; Birse, K.; Holm, J.B.; Gajer, P.; Humphrys, M.S.; Garber, D.; Guenthner, P.; Nël-Romas, L.; Abou, M.; McCorrister, S.; et al. Impact of the griffithsin anti-HIV microbiocide and placebo gels of the rectal mucose proteome and microbiome in non-human primates. *Sci. Rep.* **2018**, *8*, 8059. [CrossRef]

65. Fitzgerald, C.; Aluko, R.E.; Hossain, M.; Rai, D.K.; Hayes, M. Potential of a renin inhibitory peptide from the red seaweed *Palmaria palmata* as a functional food ingredient following confirmation and characterization of a hypotensive effect in spontaneously hypertensive rats. *J. Agric. Food Chem.* **2014**, *62*, 8352–8356. [CrossRef]

66. Sato, M.; Hosokawa, T.; Yamaguchi, T.; nakano, T.; Muramoto, K.; Kahara, T.; Funayama, K.; Kobayashi, A.; Nakano, T. Angiotensin I-Conerting enzyme inhibitory peptides derived from Wakane (*Undaria pinnatifida*) and their antihypertensive effect in spontaneously hypertensive rats. *J. Agric. Food Chem.* **2002**, *50*, 6245–6252. [CrossRef] [PubMed]

67. Suárez, Y.; González, L.; Cuadrado, A.; Berciano, M.; Lafarga, M.; Muñoz, A. Kahalalide F, a new marine-derived compound, induces oncosis in human prostate and breast cancer cells. *Mol. Cancer Ther.* **2003**, *2*, 863–872.

68. Aryee, A.N.; Agyei, D.; Akanbi, T.O. Recovery and utilization of seaweed pigments in food processing. *Curr. Opin. Food Sci.* **2018**, *19*, 113–119. [CrossRef]

69. Fernández-Rojas, B.; Hernández-Juárez, J.; Pedraza-Chaverri, J. Nutraceutical properties of phycocyanin. *J. Funct. Foods* **2014**, *11*, 375–392. [CrossRef]

70. Hao, S.; Yan, Y.; Li, S.; Zhao, L.; Zhang, C.; Liu, L.; Wang, C. The in vitro anti-tumor activity of phycocyanin against non-small cell lung cancer cells. *Mar. Drugs* **2018**, *16*, 178. [CrossRef] [PubMed]

71. Jiang, L.; Wang, Y.; Yin, Q.; Liu, G.; Liu, H.; Huang, Y.; Li, B. Phycocyanin: A potential drug for cancer treatment. *J. Cancer* **2017**, *8*, 3416–3429. [CrossRef]

72. Manirafasha, E.; Ndikubwimana, T.; Zeng, X.; Lu, Y.; Jing, K. Phycobiliprotein: Potential microalgae derived pharmaceutical and biological reagent. *Biochem. Eng. J.* **2016**, *109*, 282–296. [CrossRef]

73. Kuda, T.; Kosaka, M.; Hirano, S.; Kawahara, M.; Sato, M.; Kaneshima, T.; Nishizawa, M.; Takahashi, H.; Kimura, B. Effect of sodium-alginate and laminaran on *Salmonella* Typhimurium infection in human enterocyte-like HT-29-Luc cells and BALB/c mice. *Carbohydr. Polym.* **2015**, *125*, 113–119. [CrossRef]

74. Han, Z.; Yang, M.; Fu, X.; Chen, M.; Su, Q.; Zhao, Y.; Mou, H. Evaluation of prebiotic potential of three marine algae oligosaccharides from enzymatic hydrolysis. *Mar. Drugs* **2019**, *17*, 173. [CrossRef]

75. Costa, L.; Fidelis, G.; Cordeiro, S.L.; Oliveira, R.; Sabry, D.D.A.; Câmara, R.; Nobre, L.; Costa, M.; Almeida-Lima, J.; Farias, E. Biological activities of sulfated polysaccharides from tropical seaweeds. *Biomed. Pharmacother.* **2010**, *64*, 21–28. [CrossRef] [PubMed]

76. Zaporozhets, T.; Besednova, N.; Kuznetsova, T.; Zvyagintseva, T.; Makarenkova, I.; Kryzhanovsky, S.; Melnikov, V. The prebiotic potential of polysaccharides and extracts of seaweeds. *Russ. J. Mar. Biol.* **2014**, *40*, 1–9. [CrossRef]

77. Hall, A.B.; Tolonen, A.C.; Xavier, R.J. Human genetic variation and the gut microbiome in disease. *Nat. Rev. Genet.* **2017**, *18*, 690. [CrossRef] [PubMed]

78. Roca-Saavedra, P.; Mendez-Vilabrille, V.; Miranda, J.M.; Nebot, C.; Cardelle-Cobas, A.; Franco, C.M.; Cepeda, A. Food additives, contaminants and other minor components: Effects on human gut microbiota—A review. *J. Physiol. Biochem.* **2018**, *74*, 69–83. [CrossRef] [PubMed]

79. Sardari, R.R.; Nordberg Karlsson, E. Marine poly-and oligosaccharides as prebiotics. *J. Agric. Food Chem.* **2018**, *66*, 11544–11549. [CrossRef]

80. Huttenhower, C.; Gevers, D.; Knight, R.; Abubucker, S.; Badger, J.H.; Chinwalla, A.T.; Creasy, H.H.; Earl, A.M.; Fitzgerald, M.G.; Fulton, R.S. Structure, function and diversity of the healthy human microbiome. *Nature* **2012**, *486*, 207.

81. Zmora, N.; Suez, J.; Elinav, E. You are what you eat: Diet, health and the gut microbiota. *Nat. Rev. Gastroenterol. Hepatol.* **2019**, *16*, 35–56. [CrossRef]

82. Kong, Q.; Dong, S.; Gao, J.; Jiang, C. In vitro fermentation of sulfated polysaccharides from *E. prolifera* and *L. japonica* by human fecal microbiota. *Int. J. Biol. Macromol.* **2016**, *91*, 867–871. [CrossRef]

83. Clemente, J.C.; Pehrsson, E.C.; Blaser, M.J.; Sandhu, K.; Gao, Z.; Wang, B.; Magris, M.; Hidalgo, G.; Contreras, M.; Noya-Alarcón, Ó. The microbiome of uncontacted Amerindians. *Sci. Adv.* **2015**, *1*, e1500183. [CrossRef]

84. Teng, Z.; Qian, L.; Zhou, Y. Hypolipidemic activity of the polysaccharides from *Enteromorpha prolifera. Int. J. Biol. Macromol.* **2013**, *62*, 254–256. [CrossRef]

85. Charoensiddhi, S.; Conlon, M.A.; Vuaran, M.S.; Franco, C.M.; Zhang, W. Impact of extraction processes on prebiotic potential of the brown seaweed *Ecklonia radiata* by in vitro human gut bacteria fermentation. *J. Funct. Foods* **2016**, *24*, 221–230. [CrossRef]

86. Praveen, M.A.; Parvathy, K.K.; Jayabalan, R.; Balasubramanian, P. Dietary fiber from Indian edible seaweeds and its in-vitro prebiotic effect on the gut microbiota. *Food Hydrocoll.* **2019**, *96*, 343–353. [CrossRef]

87. Fu, X.; Cao, C.; Ren, B.; Zhang, B.; Huang, Q.; Li, C. Structural characterization and in vitro fermentation of a novel polysaccharide from *Sargassum thunbergii* and its impact on gut microbiota. *Carbohydr. Polym.* **2018**, *183*, 230–239. [CrossRef]

88. Rodrigues, D.; Walton, G.; Sousa, S.; Rocha-Santos, T.A.; Duarte, A.C.; Freitas, A.C.; Gomes, A.M. In vitro fermentation and prebiotic potential of selected extracts from seaweeds and mushrooms. *LWT-Food Sci. Technol.* **2016**, *73*, 131–139. [CrossRef]

89. Devillé, C.; Gharbi, M.; Dandrifosse, G.; Peulen, O. Study on the effects of laminarin, a polysaccharide from seaweed, on gut characteristics. *J. Sci. Food Agric.* **2007**, *87*, 1717–1725. [CrossRef]

90. Kim, J.; Yu, D.; Kim, J.; Choi, E.; Lee, C.; Hong, Y.; Kim, C.; Lee, S.; Choi, I.; Cho, K. Effects of *Undaria linnatifida* and *Laminaria japonica* on rat's intestinal microbiota and metabolite. *J. Nutr. Food Sci.* **2016**, *6*. [CrossRef]

91. Strain, C.R.; Collins, K.C.; Naughton, V.; McSorley, E.M.; Stanton, C.; Smyth, T.J.; Soler-Vila, A.; Rea, M.C.; Ross, P.R.; Cherry, P. Effects of a polysaccharide-rich extract derived from Irish-sourced *Laminaria digitata* on the composition and metabolic activity of the human gut microbiota using an in vitro colonic model. *Eur. J. Nutr.* **2019**, 1–17. [CrossRef]

92. Zhao, J.; Cheung, P.C. Fermentation of β-glucans derived from different sources by bifidobacteria: Evaluation of their bifidogenic effect. *J. Agric. Food Chem.* **2011**, *59*, 5986–5992. [CrossRef]

93. Kaewmanee, W.; Suwannaporn, P.; Huang, T.C.; Al-Ghazzewi, F.; Tester, R.F. In vivo prebiotic properties of *Ascophyllum nodosum* polysaccharide hydrolysates from lactic acid fermentation. *J. Appl. Phycol.* **2019**, *31*, 3153–3162. [CrossRef]

94. Ramnani, P.; Chitarrari, R.; Tuohy, K.; Grant, J.; Hotchkiss, S.; Philp, K.; Campbell, R.; Gill, C.; Rowland, I. In vitro fermentation and prebiotic potential of novel low molecular weight polysaccharides derived from agar and alginate seaweeds. *Anaerobe* **2012**, *18*, 1–6. [CrossRef]

95. Shang, Q.; Shan, X.; Cai, C.; Hao, J.; Li, G.; Yu, G. Dietary fucoidan modulates the gut microbiota in mice by increasing the abundance of *Lactobacillus* and Ruminococcaceae. *Food Funct.* **2016**, *7*, 3224–3232. [CrossRef]

96. Huebbe, P.; Nikolai, S.; Schloesser, A.; Herebian, D.; Campbell, G.; Gluer, C.C.; Zeyner, A.; Demetrowitsch, T.; Schwarz, K.; Metges, C.C.; et al. An extract from the Atlantic brown algae *Saccorhiza polyschides* counteracts diet-induced obesity in mice via a gut related multi-factorial mechanisms. *Oncotarget* **2017**, *8*, 73501–73515. [CrossRef] [PubMed]

97. Berri, M.; Olivier, M.; Holbert, S.; Dupont, J.; Demais, H.; Le Goff, M.; Collen, P.N. Ulvan from *Ulva armoricana* (Chlorophyta) activates the PI3K/Akt signalling pathway via TLR4 to induce intestinal cytokine production. *Algal Res.* **2017**, *28*, 39–47. [CrossRef]

98. Shih, Y.L.; Hsueh, S.C.; Chen, Y.L.; Chou, J.S.; Chung, H.Y.; Liu, K.L.; Jair, H.W.; Chuang, Y.Y.; Lu, H.F.; Liu, J.Y.; et al. Laminarin promotes immune responses and reduces lactate dehydrogenase but increases glutamic pyruvic transaminase in normal mice in vivo. *In Vivo* **2018**, *32*, 523–529. [PubMed]

99. Wang, X.; Wang, X.; Jiang, H.; Cai, C.; Li, G.; Hao, J.; Yu, G. Marine polysaccharides attenuate metabolic syndrome by fermentation products and altering gut microbiota: An overview. *Carbohydr. Polym.* **2018**, *195*, 601–612. [CrossRef]

100. Nakata, T.; Kyoui, D.; Takahashi, H.; Kimura, B.; Kuda, T. Inhibitory effects of laminaran and alginate on production of putrefactive compounds from soy protein by intestinal microbiota in vitro and in rats. *Carbohydr. Polym.* **2016**, *143*, 61–69. [CrossRef]

101. An, C.; Kuda, T.; Yazaki, T.; Takahashi, H.; Kimura, B. FLX pyrosequencing analysis of the effects of the brown-algal fermentable polysaccharides alginate and laminaran on rat cecal microbiotas. *Appl. Environ. Microbiol.* **2013**, *79*, 860–866. [CrossRef]

102. Xie, G.; Wang, X.; Liu, P.; Wei, R.; Chen, W.; Rajani, C.; Hernandez, B.Y.; Alegado, R.; Dong, B.; Li, D.; et al. Distinctly altered gut microbiota in the progression of liver disease. *Oncotarget* **2016**, *7*, 19355–19366. [CrossRef]

103. Scott, K.P.; Martin, J.C.; Duncan, S.H.; Flint, H.J. Prebiotic stimulation of human colonic butyrate-producing bacteria and bifidobacteria, in vitro. *FEMS Microbiol. Ecol.* **2014**, *87*, 30–40. [CrossRef]

104. Zheng, C.; Liu, R.; Xue, B.; Luo, J.; Gao, L.; Wang, Y.; Ou, S.; Li, S.; Peng, X. Impact and consequences of polyphenols and fructooligosaccharide interplay on gut microbiota in rats. *Food Funct.* **2017**, *8*, 1925–1932. [CrossRef]

105. Di, T.; Chen, G.; Sun, Y.; Ou, S.; Zeng, X.; Ye, H. In vitro digestion by saliva, simulated gastric and small intestinal juices and fermentation by human fecal microbiota of sulfated polysaccharides from *Gracilaria rubra*. *J. Funct. Foods* **2018**, *40*, 18–27. [CrossRef]

106. Silva, R.; Santana, A.; Carvalho, N.; Bezerra, T.; Oliveira, C.; Damasceno, S.; Chaves, L.; Freitas, A.; Soares, P.; Souza, M. A sulfated-polysaccharide fraction from seaweed *Gracilaria birdiae* prevents naproxen-induced gastrointestinal damage in rats. *Mar. Drugs* **2012**, *10*, 2618–2633. [CrossRef] [PubMed]

107. Bajury, D.M.; Rawi, M.H.; Sazali, I.H.; Abdullah, A.; Sarbini, S.R. Prebiotic evaluation of red seaweed (*Kappaphycus alvarezii*) using in vitro colon model. *Int. J. Food Sci. Nutr.* **2017**, *68*, 821–828. [CrossRef] [PubMed]

108. Fu, X.T.; Kim, S.M. Agarase: Review of major sources, categories, purification method, enzyme characteristics and applications. *Mar. Drugs* **2010**, *8*, 200–218. [CrossRef] [PubMed]

109. Shang, Q.; Jiang, H.; Cai, C.; Hao, J.; Li, G.; Yu, G. Gut microbiota fermentation of marine polysaccharides and its effects on intestinal ecology: An overview. *Carbohydr. Polym.* **2018**, *179*, 173–185. [CrossRef]

110. Zhang, N.; Mao, X.; Li, R.W.; Hou, E.; Wang, Y.; Xue, C.; Tang, Q. Neoagarotetraose protects mice against intense exercise-induced fatigue damage by modulating gut microbial composition and function. *Mol. Nutr. Food Res.* **2017**, *61*, 1600585. [CrossRef]

111. Ladirat, S.; Schols, H.; Nauta, A.; Schoterman, M.; Schuren, F.; Gruppen, H. In vitro fermentation of galacto-oligosaccharides and its specific size-fractions using non-treated and amoxicillin-treated human inoculum. *Bioact. Carbohydr. Diet. Fibre* **2014**, *3*, 59–70. [CrossRef]

112. Kulshreshtha, G.; Rathgeber, B.; Stratton, G.; Thomas, N.; Evans, F.; Critchley, A.; Hafting, J.; Prithiviraj, B. Feed supplementation with red seaweeds, *Chondrus crispus* and *Sarcodiotheca gaudichaudii*, affects performance, egg quality, and gut microbiota of layer hens. *Poult. Sci.* **2014**, *93*, 2991–3001. [CrossRef]

113. Munoz-Munoz, J.; Cartmell, A.; Terrapon, N.; Henrissat, B.; Gilbert, H.J. Unusual active site location and catalytic apparatus in a glycoside hydrolase family. *Proc. Natl. Acad. Sci. USA* **2017**, *114*, 4936–4941. [CrossRef]

114. Ren, X.; Liu, L.; Gamallat, Y.; Zhang, B.; Xin, Y. *Enteromorpha* and polysaccharides from *Enteromorpha* ameliorate loperamide-induced constipation in mice. *Biomed. Pharmacother.* **2017**, *96*, 1075–1081. [CrossRef]

115. Zhang, Z.; Wang, X.; Han, S.; Liu, C.; Liu, F. Effect of two seaweed polysaccharides on intestinal microbiota in mice evaluated by illumina PE250 sequencing. *Int. J. Biol. Macromol.* **2018**, *112*, 796–802. [CrossRef] [PubMed]

116. Rahimi, F.; Tabarsa, M.; Rezaei, M. Ulvan from green algae *Ulva intestinalis*: Optimization of ultrasound-assisted extraction and antioxidant activity. *J. Appl. Phycol.* **2016**, *28*, 2979–2990. [CrossRef]

117. Lecerf, J.; Dépeint, F.; Clerc, E.; Dugenet, Y.; Niamba, C.N.; Rhazi, L.; Cayzeele, A.; Abdelnour, G.; Jaruga, A.; Younes, H. Xylo-oligosaccharide (XOS) in combination with inulin modulates both the intestinal environment and immune status in healthy subjects, while XOS alone only shows prebiotic properties. *Br. J. Nutr.* **2012**, *108*, 1847–1858. [CrossRef] [PubMed]

118. Lahaye, M.; Michel, C.; Barry, J.L. Chemical, physicochemical and in-vitro fermentation characteristics of dietary fibres from *Palmaria palmata* (L.) Kuntze. *Food Chem.* **1993**, *47*, 29–36. [CrossRef]

Review

Avocado Oil: Characteristics, Properties, and Applications

Marcos Flores [1,*], **Carolina Saravia** [2,*], **Claudia E. Vergara** [1], **Felipe Avila** [3], **Hugo Valdés** [4] and **Jaime Ortiz-Viedma** [5,*]

[1] Departamento de Ciencias Básicas, Facultad de Ciencias, Universidad Santo Tomás, Avenida Carlos Schorr 255, Talca 3473620, Chile; claudiavergarava@santotomas.cl

[2] Escuela de Nutrición y Dietética, Facultad de Salud, Universidad Santo Tomás, Avenida Carlos Schorr 255, Talca 3473620, Chile

[3] Escuela de Nutrición y Dietética, Facultad de Ciencias de la Salud, Universidad de Talca, Talca 3460000, Chile; favilac@utalca.cl

[4] Centro de Innovación en Ingeniería Aplicada, Departamento de Computación e Industrias, Universidad Catolica del Maule, Avenida San Miguel 3605, Talca 3480112, Chile; hvaldes@ucm.cl

[5] Department of Food Science and Chemical Technology, Faculty of Chemical and Pharmaceutical Sciences, University of Chile, Santos Dumont 964, Santiago 8380494, Chile

* Correspondence: marcosflores@santotomas.cl (M.F.); csaraviam@santotomas.cl (C.S.); jaortiz@uchile.cl (J.O.-V.); Tel.: +56-7123-424-18 (M.F.)

Academic Editor: Jose M. Miranda
Received: 9 May 2019; Accepted: 8 June 2019; Published: 10 June 2019

Abstract: Avocado oil has generated growing interest among consumers due to its nutritional and technological characteristics, which is evidenced by an increase in the number of scientific articles that have been published on it. The purpose of the present research was to discuss the extraction methods, chemical composition, and various applications of avocado oil in the food and medicine industries. Our research was carried out through a systematic search in scientific databases. Even though there are no international regulations concerning the quality of avocado oil, some authors refer to the parameters used for olive oil, as stated by the Codex Alimentarius or the International Olive Oil Council. They indicate that the quality of avocado oil will depend on the quality and maturity of the fruit and the extraction technique in relation to temperature, solvents, and conservation. While the avocado fruit has been widely studied, there is a lack of knowledge about avocado oil and the potential health effects of consuming it. On the basis of the available data, avocado oil has established itself as an oil that has a very good nutritional value at low and high temperatures, with multiple technological applications that can be exploited for the benefit of its producers.

Keywords: avocado oil; oil extraction; antioxidants compounds; fatty acid profile

---

1. Introduction

Avocado (*Persea americana Mill.*) is a fruit native to Central America, grown in warm temperate and subtropical climates throughout the world. The pulp of this fruit contains about 60% oil, 7% skin, and approximately 2% seed [1,2]. The main producers of avocado oil in the world are New Zealand, Mexico, the United States, South Africa, and Chile [3]. Avocado oil has sparked a growing interest in human nutrition, food industry, and cosmetics. The lipid content, mainly of monounsaturated fatty acids, is associated with cardiovascular system benefits and anti-inflammatory effects [4,5].

There are no internationally defined parameters for avocado oil. The values that are commonly used are those recommended for olive oil. The quality standard for olive oil is available in the Codex Alimentarius and the International Olive Oil Council (IOC) [6].

Woolf et al. [7] proposed a classification for avocado oil based on its extraction method and fruit quality. Avocado oil of a higher quality, "extra virgin", corresponds to that produced from high-quality fruit, extracted only with mechanical methods, using a temperature below 50 °C and without the use of chemical solvents. "Virgin" avocado oil is produced with fruit of a lower quality (with small areas of rot and physical alterations), extracted by mechanical methods, using a temperature below 50 ° C and without the use of chemical solvents. "Pure" avocado oil is a type of oil for the production of which the quality of the fruit is not important; it is a bleached and deodorized oil, infused with the natural flavor of herbs or fruits. Finally, "mixed" avocado oil is combined with olive, macadamia, and other oils. Therefore, it presents sensory and chemical characteristics that are variable.

The Mexican norm [8] states that the "crude oil of avocado" is a slightly amber-colored fatty liquid, obtained by physical extraction of the pulp and the seed of the fruit (*Persea americana*). "Pure" edible avocado oil is a product with at least 98.5% refined avocado oil.

In this work, a systematic review of the literature was carried out to collect, select, evaluate, and summarize all available evidence regarding the processes and properties of avocado oil. The research question was: What are the most published topics on avocado oil? The answer to this question allowed us to include topics, such as: extraction methods (i.e., cold pressed method, ultrasound-assisted aqueous extraction method, supercritical CO_2 method, CO_2 subcritical method, enzymatic extraction, and solvent extraction), procedures of conservation, contamination/adulteration, technological applications, composition (characteristics according to the variety and origin of the fruit, physicochemical characterization, avocado seed oil, and comparison with other oils), and biological effects (human health effects and experimental studies in animals). Figure 1 shows the exponential increase of scientific interest in avocado oil. On the topic, "avocado oil" (from 1980 to date), 180 and 224 articles have been published in the Web of Science (WoS) and Scopus, respectively.

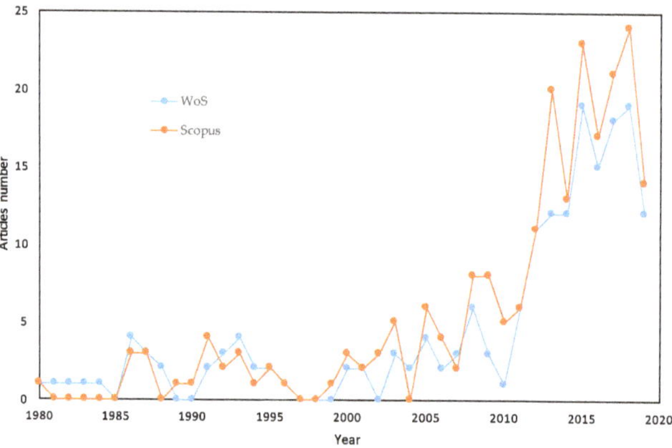

Figure 1. Graphic of articles (topic: "avocado oil") from 1980 to 2019 in the Web of Science (WoS) and Scopus.

The purpose of this research is to produce a complete profile on avocado oil, including extraction methods, physicochemical characteristics, nutritional properties, as well as various applications in the food and medicine industries. Avocado oil has proven to be a vegetable oil with a composition of major and minor components that are highly appreciated by the population, either at low or high temperatures, with multiple technological applications.

2. Extraction Methods for Avocado Oil

The information presented in this section focuses on the process efficiency, improvement of production performance, and product quality as well as applications in the food industry.

Considering the high humidity percentage of avocado (around 70 to 80%), the influence of the pulp drying method prior to oil extraction has been studied [9]. The quality parameters (peroxide value, iodine value, amount of oleic acid, refractive index, electrical conductivity, content of carotenoids, chlorophyll, phenolic compounds, and antioxidant activity) have shown better results when the pulp is dried at 60 °C under vacuum, and the extraction is performed by the Soxhlet method. Meanwhile, the bioactive compounds were best preserved when the avocado pulp was dried at 60 °C with air ventilation and mechanical pressing [10]. On the other hand, avocado pulp oil, pressed and dried in a microwave, presented a better quality—determined by the acidity index, peroxide index, and oxidative stability—when compared with oil obtained by extraction with ethanol. The composition of fatty acids did not differ significantly when analyzing oil obtained by drying under microwaves or in a drying oven with forced air circulation [11]. According to Chimsook and Assawarachan [12], the studied drying method of the avocado pulp, prior to the extraction of the oil, does not significantly influence the composition of fatty acids. However, changes were determined in the antioxidant activity and vitamin E content of cold-pressed avocado oil (from Thailand). Higher antioxidant activities and a higher vitamin E content were observed in oil, the pulp of which was dried with hot air, when compared to oils obtained by an air-dried and vacuum process. This study is consistent with the fact that oils from the *fortune* avocado variety, obtained by the pulp drying lyophilization method, resulted in lower concentrations of α-tocopherol, squalene and β-sitosterol, as well as higher relative concentrations of campesterol and cycloartenol acetate, compared to oils obtained through hot air-drying processes [13].

2.1. Cold Pressed Method

According to the CODEX STAN 19-1981 [14], the method of extraction of edible vegetable oils is characterized by mechanical procedures, for example, extrusion or pressing, without the application of heat. In addition, the oil can only be purified by washing, sedimentation, filtration, and centrifugation.

In the cold pressing method, oil recovery is only obtained from the parenchyma cells of the pulp; its rupture begins in the first stages of grinding and it can be seen that the idioblastic cells (oil carriers) remain intact during the process of extraction. The extraction yield increases when the pulp is beaten at 45.5 °C for 2 h [15]. In this method, a lower extraction yield is obtained, although with higher concentrations of α-tocopherol and squalene, as well as lower contents of campesterol and cycloartenol acetate, compared to the Soxhlet method [13]. Drying by lyophilization and subsequent extraction by the Soxhlet method allows for a better extraction performance. However, when drying by lyophilization and extracting by cold pressing, oils with a greater concentration of antioxidants, and other bioactive compounds were obtained [13].

2.2. Ultrasound-Assisted Aqueous Extraction Method (UAAE)

This method uses the cavitation forces produced by acoustic waves to break down the cell walls of the oil-containing cells. This process allows for the generation of an emulsion, which facilitates oil extraction. This method can be carried out using an ultrasonic bath or an ultrasonic horn transducer [16]. The high frequency ultrasound conditioning (0.4, 0.6, and 2 MHz, 5 min, 90 kJ/kg) of the avocado puree can improve the oil separation and potentially reduce the beating time in industrial processes, without affecting the quality of the oil. If this treatment is applied after shaking, the extractability of the oil increases by between 2% and 5%. The oils obtained from sonicated purees showed free fatty acids (FFA) and peroxide values below the levels of industrial specification (peroxide less than 20 meqO$_2$/kg) and an increase in total phenolic compounds after a 2 MHz treatment [17]. The ultrasound-assisted aqueous extraction (UAAE) of low virgin avocado oil in FFA, considered as virgin avocado oil, is that obtained by mechanical or natural means at low temperatures (<50 °C) and without chemical refining [7]. The optimal UAAE parameters to produce the highest extraction of virgin avocado oil was 6 mL/g water-dried pulp powder, 30 min of sonication time at 35 °C. The sonicated virgin avocado oil was lighter and had a higher level of unsaturated fatty acids, compared to the avocado oil extracted by the Soxhlet method [16].

2.3. Supercritical CO_2 Method

This method of extraction is based on the use of supercritical fluids, substances that are, in certain circumstances, in a state in which they have intermediate properties between liquid and gas. Supercritical CO_2 ($scCO_2$) is a totally innocuous gas, which becomes a powerful solvent under conditions of pressure and at a temperature above its critical point [18].

Extraction with $scCO_2$ presents a higher performance at a pressure of 400 bar. The use of ethanol as a co-solvent favors the extraction of residual oil, benefiting the extraction of a fraction enriched in tocopherols [19].

Some authors [20,21] proposed the combined extraction of avocado oil and active compounds present in peppers (capsanthin) and tomatoes (lycopene) using $scCO_2$ in order to enrich the avocado oil. For this, a fixed bed extractor was used, where the lipids and the desired active ingredient were subjected to the extraction process, simultaneously with $scCO_2$. First, both the $scCO_2$ and the oil extracted from the avocado passed through the avocado bed and then through the second bed, where the plant component to be co-extracted was found. The lipids obtained in the first chamber served as a co-solvent with $scCO_2$ for extraction in the second chamber. In the case of the simultaneous extraction of edible avocado oil and the capsanthin (carotenoid) of red pepper, the higher concentration of oil improved the extraction yield of capsanthin. However, a less concentrated extract was obtained, since the carotenoid was diluted in the product. In the case of the extraction of avocado oil rich in lycopene, the extraction yield of lycopene increased as the proportion of avocado in the first extraction chamber increased, being the best condition for the extraction of lycopene present in tomato pomace at 400 bar and 50 °C.

Restrepo et al. [22] evaluated the quality of avocado oil extracted by Soxhlet, cold pressed, and $scCO_2$ methods, determining the quality of the oil in terms of free fatty acid titration, peroxide index, iodine index, saponification, and specific gravity, according to the American Oil Chemists' Society (AOCS) standards. Extraction with supercritical fluids was the technique by which the highest yields and quality were obtained. Oils extracted by $scCO_2$ were characterized as possessing a lower acidity index (0.48%), low oxidation of unsaturated fatty acids (16.87 meqO$_2$/kg) and higher iodine index (80.18 cgI2/g), when compared with the other methods. In addition, extraction by cold pressing showed better results in terms of vitamin E content.

Regarding extraction by pressurized fluids, the extraction with the liquefied gas of compressed oil (LPG), constituted by a mixture of propane, n-butane, isobutane, ethane, and other hydrocarbons, showed a higher oil extraction performance in less time and with a lower solvent consumption than the $scCO_2$ method. On the other hand, the oil obtained by compressed LPG presented higher concentrations of Stigmasterol, licopersene, palmitic acid, oleic acid, and linoleic acid. However, $scCO_2$ provided a higher yield in terms of antioxidant activity, which was determined by means of the 2,2-Diphenyl-1-picrylhydrazyl (DPPH) radical assay [23].

2.4. CO_2 Subcritical Method

Extraction with sCO_2 operates under the same principle as the $scCO_2$ extraction, but with a temperature below 31.1 °C and CO_2 pressure of 72.9 bar [24].

In this part of the review, a comparison of the physicochemical properties of avocado oil, extracted through sCO_2, UAAE, and conventional solvent (AOAC 920.39 [25]) will be analyzed. Extraction with sCO_2 was performed at 27 °C and 68 bar CO_2, UAAE was performed with 60 mL of distilled water, an ultrasonic power of 240 W and a frequency of 40 kHz for 30 min at 35 °C, followed by a final pressing. Compared to solvent extraction, oils extracted using sCO_2 and UAAE had higher iodine index values but lower melting points, determined by slip, free fatty acid content, and saponification index values. The oils extracted by sCO_2 and UAAE have a clear color and higher levels of unsaturated fatty acids than the oil extracted with hexane. Regardless of the extraction method, the main fatty acids in avocado oils were oleic and palmitic acids, while the main triacylglycerols in avocado oils were palmitoyl-dioleoyl-glycerol (POO; 22.48–23.01%) and palmitoyl- Oleoyl-linoleoyl-glycerol (POL; 17.64–18.23%) [24].

2.5. Enzymatic Extraction

In order to improve the performance of extraction by centrifugation, the incorporation of enzymes, such as pectinases, α-amylase, proteases, and cellulase, to avocado paste have been considered. The yield varies depending on the concentration and type of enzyme used and the reaction time and percentage of water used. It is emphasized that this method improves oil by up to 25 times, in comparison with the performance of a non-enzymatic centrifugation [26].

2.6. Solvent Extraction

Reddy et al. [27] compared four extraction methods to produce avocado oil (*Hass* and *Fuerte* variety). They analyzed: (1) the extraction with traditional solvent using Soxhlet (5.0 g of dried avocado sample with 250 mL of hexane for 24 h); (2) Ultrasonic Soxhlet extraction (5.0 g of dry avocado sample, sonicated in a water bath at 60 °C, with hexane as the solvent, for 1 h); (3) Soxhlet extraction, combined with a microwave treatment (avocado paste 5 mm thick, extended in the rotating plate of a domestic microwave oven, heated to the maximum power for 11 min, with 5 g of the resulting mass subsequently extracted by means of the Soxhlet method with hexane); and (4) extraction with supercritical fluid (Argon and scCO$_2$ used as extraction fluids, extractions performed for 2 h, with a fluid flow rate of 2.8–3.5 mL/min). The traditional Soxhlet extraction method yields the most reproducible results, whereas the microwave extraction showed a higher extraction yield and higher fatty acid content (69.94%).

Meyer and Terry [28] performed a sequential extraction and quantification of fatty acids and avocado sugars. The average oil yield using Soxhlet extraction, with ethanol as the solvent, was significantly higher than the oil obtained by homogenization with hexane, and the fatty acid profiles for the two methods were similar. As the maturity of the fruit increases, the extraction of oil is improved. After lipid removal, methanolic extraction was superior in terms of the sucrose and perseitol obtained, compared to extraction with 80% ethanol (*v/v*). The extraction of mannoheptulose was not affected by any of the solvents used.

The yield of avocado oil extraction has been assessed, comparing four extraction methods using solvents of different polarities. The extraction was performed using the Soxhlet method, with (1) petroleum ether, (2) homogenization with petroleum ether, (3) homogenization with a mixture of chloroform/methanol (2:1 *v/v*), and finally (4) extraction with chlorine-naphthalene and ball milling. It was determined that methods that only use petroleum ether as an extraction medium presented lower yields (6–9% less) than the last two methods. Saponifiable residues were lower when the method using the chloroform/methanol mixture was employed. However, this method did not completely eliminate the residual oil from the fruit [29].

Ortiz-Moreno et al. [30] analyzed the effect of four extraction methods on the chemical-physical quality of avocado oil, namely, (1) the method of microwave extraction and manual pressing, (2) extraction with hexane using the Soxhlet methodology, (3) microwave extraction, combined with the Soxhlet methodology, using hexane as the solvent, and (4) extraction with acetone. The method with the highest oil extraction performance was the third method. The amount of *trans* fatty acids produced by the first method was the lowest and the latter method is also the one that generates the least physicochemical alterations.

When analyzing the effect of the drying method and avocado oil extraction process, ripe fruit, independent of the drying method, presents a higher extraction performance than immature fruit. This is influenced by the enzymatic degradation of the cell wall of the parenchyma during maturation. Freeze drying improves the amount of oil extracted for the scCO$_2$ extracts and, to a lesser extent, for the hexane extracts. Extraction with hexane has been shown to have a higher oil extraction yield than scCO$_2$ due to the lower degree of selectivity of this solvent, which completely penetrates the plant material [31].

Regarding the performance of the avocado oil extraction process, methodologies have been proposed for developing countries. One is carried out with boiling petroleum ether (30–60 °C) and another extraction with distilled water (avocado paste, diluted at a ratio of 3:1 and 5:1 (*w/w*), heated in a water bath of 75 to 98 °C, with subsequent centrifugation. In the extraction methods, calcium chloride, sodium chloride, calcium carbonate and calcium sulfate were used as extraction aids. The presence of inorganic salts at a low concentration improves the extraction performance, provided that it does not exceed 5%; otherwise, it has an adverse effect. The most efficient extractions were obtained with a water/avocado ratio of 5:1, pH of 5.5 and centrifugal force of 12,300× *g*, with the addition of 5% calcium carbonate or calcium sulfate. At higher heating temperatures (75–98 °C), the oil release time decrease. In addition, the gravity sedimentation for four days at 37 °C, followed by centrifugation, improves the oil extraction performance [32].

Considering the use of organic solvents in avocado oil extraction processes could alter the quality of fatty acids by inducing the formation of *trans* isomers. Ariza-Ortega et al. [33] proposed the application of infrared spectroscopy by Fourier transform (FTIR) to study trans fatty acids in the avocado oils of the *Hass*, *Fuerte*, and *Criollo* varieties. For this, oil extraction was performed by centrifugation at 40 °C and extraction with hexane at 70 °C for 4 h. The method using centrifugation did not increase the deterioration of fatty acids. A strong band at 723 cm^{-1} was documented, which is attributable to the cis functional groups, where the green color was maintained. On the other hand, the infrared spectroscopy with Fourier transform (FTIR) analysis identified an absorption band, located at 968 cm^{-1}, which is associated with fatty acids, with *trans* isomerism for the *Fuerte* variety extracted with hexane.

3. Procedures for the Conservation of Avocado Oil

The conservation of oils is a necessary issue to address, since it allows for increasing the useful life of the products. One of the efforts made to improve the conservation of avocado oil has been the use of physical techniques, such as the electric field.

The electric field (voltage 9 kV cm^{-1}, frequency 720 Hz, time of 5 and 25 min) allows the polyphenol oxidase enzyme present in the avocado pulp to be inactivated, preserving the components present in the avocado oil. The modifications in the quality of the refined oil (established according to the acidity index, peroxides, and iodine) are minimal, considering the electric field method as an alternative for the addition of synthetic antioxidants [34].

The oxidative stability (determined by finding the antioxidant activity reducing ferric ion, FRAP), during the storage of cold-extracted avocado oil in the presence of the oleoresins of *Capsicum annuum* L. (vegetable material rich in carotenoids), was assessed. It was determined that the optimal extraction of carotenoids was at a concentration of 1:3 (*w/v*: *Capsicum annuum* L/avocado oil) for 48 h in darkness at room temperature. The behavior of the oil under stronger conditions (45 °C, 30 days) showed the following characteristics: (1) the extracts were stable to lipid oxidation, with a Totox index total value of 27.34, (2) 85.6% of carotenoids were conserved, (3) 80.66% of the antioxidant activity was retained, and (4) there was a color change (ΔE) of 1.783. The oleoresins obtained by extraction with avocado oil can be considered as an economic and sustainable alternative for the extraction of carotenoids, with a good oxidative stability, compared with organic solvents [35].

4. Use of Analytical Techniques in the Quantification, Adulteration, and Contamination of Avocado Oil

There is growing interest among consumers in accessing quality and authentic products. Vegetable oils can suffer from contamination and/or adulteration, which causes the product to have components not specific to the oil. It is here that the development, implementation, and application of analytical technologies are very useful.

The components present in avocado oil, such as fatty acids and phytosterols, have been quantified, mainly by gas chromatography, coupled with a flame ionization detector (GC-FID). In addition, techniques, such as ultra-high-performance liquid chromatography (UHPLC), coupled with mass spectrometry (UHPLC-MS) or a photodiode array detector (UHPLC-PDA), as well as Inductively Coupled Plasma Mass Spectrometry (ICP-MS), have been used for the identification and/or quantification of analytes, such as polyphenols, squalene and minerals, respectively [36,37]. Other analytical techniques have been used for the qualitative determination of the components present in avocado oil, including ^{13}C nuclear magnetic resonance spectroscopy (NMR), which has been used for the identification of its major components, including fatty acids [38]. At the same time, ^{1}H Nuclear magnetic resonance spectroscopy (^{1}H-NMR) has been used for the detection of the minor components present in other vegetable oils [39]. Therefore, the development of new analytical methodologies for quantifying the analytes present in avocado oil represents a major challenge.

Rohman et al. [40] studied the purity of avocado oil, adulterated with palm oil and canola oil, through FTIR, combined with chemometric techniques. FTIR combined with multivariate calibrations can be used to detect and quantify the adulteration of avocado oil in binary mixtures with palm oil and canola oil.

The adulteration of avocado oil with soybean oil or grape seed oil can be determined using mid-infrared spectroscopy, combined with the statistical method of partial least squares discriminant analysis. This methodology allows for a simple and fast discrimination of avocado oil in binary mixtures and Tertiary oils. The frequency selected for the authentication of avocado oil was 1500–750 cm^{-1}, with a precision of 100% for the analysis of the mixture of two oils and 93.3% for the mixture of three oils [41].

Organophosphorus pesticides in samples of commercial avocado oil were determined using atmospheric pressure microwave-assisted liquid–liquid extraction (APMAE), with solid-phase extraction or low-temperature precipitation, as the clean-up step. The analysis was carried out by gas chromatography-flame photometric detection and gas chromatography-tandem mass spectrometry. Chlorpyrifos residues were detected in one of four samples of commercially packaged avocado oil, produced in Chile [42]. While spectroscopic techniques have focused on determining the adulteration of avocado oil with the presence of other types of vegetable oil, according to the literature reviewed here, there is a research deficiency related to the modification of the composition of avocado oil, including the study of its major components, such as triacylglycerides and/or fatty acids, in addition to its minority components, such as phytosterols, alkanes, aliphatic alcohols, polyphenols, and others. This could provide information for detecting the contamination of avocado oil with other oils of a different quality.

5. Technological Applications of Avocado Oil

At the industrial level, there is a constant demand for the production of healthy foods that can maintain their nutritional properties over time, as well as environmentally friendly technological solutions. Avocado oil is mainly sold for direct consumption due to its interesting contribution of fatty acids, vitamins, antioxidants, among other compounds. Efforts have been made to develop products based on avocado oil. Arancibia et al. [43] propose the development of O/W nanoemulsions using the natural emulsifiers, lecithin and synthetic tween 80, systems that improve the characteristics with respect to traditional emulsions, such as (i) increased dispersibility of water in the encapsulated oils, which generates slightly turbid emulsions and an easy production, and (ii) a good physical and chemical stability, as well as a high bioavailability of its lipid components.

Another interesting technological application for avocado oil has been the production of structured lipids. Caballero et al. [44] propose the elaboration of triacylglycerides of the MLM type, using regio-specific immobilized commercial lipases *sn-1.3*, where M corresponds to saturated medium-chain fatty acids (6–12 carbon atoms) at positions *sn1* and *sn3* of glycerol. L corresponds to saturated or unsaturated long-chain fatty acids (14–24 carbon atoms) in the *sn2* position. The increased interest in this type of lipids is due to the low caloric intake (average caloric density for this family of lipids 5 kcal/g). According to the literature, there are no negative effects associated with the ingestion of MLM lipids for both animals and humans.

Finally, avocado oil has also been used in the production of biodegradable polymers. Polyhydroxyalkanoates (PHAs) are linear polyesters, produced by a large number of bacteria under stress conditions, with different thermal and mechanical properties, which depend on their molecular structure. Flores-Sanchez et al. [45] prepared PHAs through a fermentative process using the bacterium *C. necátor* H-16 with avocado oil and fructose, as a carbon source. The highest yield in obtaining polymers was obtained when the addition of avocado oil was 20% *v/v*, which demonstrates the feasibility of using this oil as a renewable carbon source for the PHA production process.

6. Composition of Avocado Oil

There is a growing interest in avocado oil, including the determination of the composition of major and minor components. Therefore, for a total understanding of the nutritional and functional properties that this oil presents, it is important to consider the different varieties and parts of the fruit.

6.1. Characteristics According to the Variety and Origin of the Fruit

Avocado is a fruit grown mainly in warm temperate and subtropical climates throughout the world, so it is interesting to study how the climate and country of origin can affect the fruit quality and therefore, the oil. Thus, the oil from the fruit of the *Hass* variety, originating from crops from Mexico, Australia, the United States, and New Zealand, was characterized by a high content of 62% lipids, of which oleic (42–51%) and palmitic (20–25%) lipids were present in a greater proportion. Among the predominant triacylglycerols were OOO (21–34%) and OOP (19–24%), where O and P denote oleic and palmitic acids, respectively. On the other hand, *Hass* avocado oil from New Zealand contained a significant amount of natural pigments and unsaturated compounds, compared to oils from Mexico, Australia, and the United States [1].

Studies carried out in South America on the analysis and characterization of avocado oil showed a high content of monounsaturated fatty acids (69.4%) and a lower amount of polyunsaturated and saturated fatty acids, which were 16.6% and 14%, respectively. These studies indicate that avocado oil has a thermal stability close to 176 °C and has a lower concentration of total phenolic compounds than olive oil. Despite this, the antioxidant activity of avocado oil is similar to that of olive oil. Olive oil has a high concentration of polyphenols, such as tyrosol and hydroxytyrosol [46].

Galvão et al. [47] analyzed the fatty acid composition of the pulp, seed, and skin oil of the *Fortuna*, *Collinson*, and *Barker* varieties, indicating that there was a small variation in the composition of monounsaturated fatty acids in the skin oil among the cultivars. However, the seed oil of the Collinson variety was the best due to the lower SFA content. The SFA content for pulp oil corresponded to 22.3, 29.4 and 41.3% in the *Fortuna*, *Collinson*, and *Barker* varieties, respectively. In this sense, it was possible to affirm that the pulp oil of the *Fortuna* and *Collinson* varieties presented a better quality, in terms of fatty acid profile, than the Barker variety.

In Mexico, avocado oil from six local creole varieties (*BTancitaro, Irapuato, Orgánico, Puerto, San José*, and *STancitaro*) were analyzed and compared with oil from the *Hass* variety. It was observed that the Mexican creole genotypes had a greater thermal stability, properties resistant to oxidation, and a greater phenolic content, in comparison with the commercial oil from the *Hass* variety. In addition, these varieties showed intense fluorescent peaks at 675 and 720 nm, as well as broad absorption bands centered at 465 and 510 nm, which can be used as an identification parameter for these oils [48].

Yanty et al. [49] indicated that the avocado oil originating from three Malaysian varieties was found in a significantly lower proportion than in the Australian *Hass* variety. In addition to being in a semi-solid form, all these oils had a higher proportion of oleic acid, although they also had different proportions of palmitic and linoleic acids. Regarding the composition of TAG for local varieties, the highest was POO, followed by POL, OOO, and PPO, while in the *Hass* variety, the distribution was OOO, followed by PPO, OOL, and POL. As a result of these different compositions in TAG, differences were found in the iodine index, melting point by slip and melting and solidification characteristics.

6.2. Physicochemical Characterization

Table 1 shows the composition of the common fatty acids of the oils from different varieties and origin of avocados, discussed in this work. However, only some varieties have the C6:0, C7:0, C8:0, C9:0, C10:0, C11:0, C12:0, C13:0, C14: 0, C14:1, C15:0, C15:1, C16:1, C17:0, C17:1, C19:0, C20:0, C20:1, C20:3, C20:4, C22:0, C22:1, C22:2, C23:0, and C24:0 fatty acids in a low proportion, ranging from traces (<0.06%)–3.58%. Dreher and Davenport [50] refer to the fact that the oil coming from the *Hass* variety can contain up to 71% of monounsaturated fatty acids (MUFA), 13% of polyunsaturated fatty acids (PUFA), and 16% of saturated fatty acids (SFA).

Table 1. Composition (%) of the common fatty acids of avocado oil.

Varieties and/or Country of Origin	Palmitic 16:0	Estearic 18:0	Palmitoleic 16:1 Ω7	Oleic 18:1 Ω9	Linoleic 18:2 Ω6	α linolenic 18:3 Ω3	Ref.
	18,62	0,49	8,47	60,17	10,97	0,98	[51]
	13.4	0.6	3.9	65.3	15.2	1.3	[46]
HASS	17.37 ± 0.0015	0.63 ± 0.0002	7.52 ± 0.0002	62.89 ± 0.0019	10.64 ± 0.0004	0.72 ± 0.0001	[37]
	18.17 ± 0.02	0.37 ± 0.00	4.03 ± 0.01	51.76 ± 0.04	11.12 ± 0.01	0.59 ± 0.00	[52]
	13.7 ± 1.5	-	3.4 ± 0.4	67.4 ± 3.0	14.4 ± 1.8	1.1 ± 0.01	[53] [3]
Australia	25.63 ± 0.11	0.45 ± 0.16	7.29 ± 0.05	42.59 ± 0.16	20.87 ± 0.10	3.19 ± 0.06	
México	22.59 ± 0.23	0.24 ± 0.02	11.63 ± 0.13	49.19 ± 0.57	14.72 ± 0.06	1.63 ± 0.16	[1]
New Zealand	20.61 ± 0.16	0.30 ± 0.01	10.31 ± 0.03	50.97 ± 0.30	16.10 ± 0.11	1.72 ± 0.02	
United States	22.24 ± 0.05	0.93 ± 0.08	13.14 ± 0.01	47.69 ± 0.03	14.47 ± 0.01	1.54 ± 0.00	
FORTUNE	10. 75	0.48	3.14	74.32	10.03	0.85	[51]
	20.5	0.5	6.8	60.6	13.2	-	[47]
BACON	12.16 ± 0.04	0.38 ± 0.01	6.57 ± 0.01	61.72 ± 0.30	8.30 ± 0.02	0.44 ± 0.00	[52]
PINKERTON	16.93 ± 0.03	0.43 ± 0.01	7.33 ± 0.05	57.39 ± 0.18	8.25 ± 0.02	0.56 ± 0.00	[52]
MARGARIDA	23.66		3.58	47.20	13.46	1.60	[54]
FUERTE	21.312 ± 0.550	0.762 ± 0.021	2.391 ± 0.188	64.436 ± 0.666	9.147 ± 0.030	0.467 ± 0.016	[55]
	12.37 ± 0.01	0.51 ± 0.01	7.58 ± 0.00	64.62 ± 0.20	8.46 ± 0.02	0.47 ± 0.00	[52]
BREDA	19.9–21.3	-	2.7–7.0	57.1–64.5	10.6–11.0	0.4–0.6	[10] [1]
CRIOLLA MEXICANA	28.12–34.48	0.23–1.07	6.64–8.5	40.73–42.72	15.52–18.88	1.51–2.14	[24] [2]
DE MALASIA	30.37 ± 0.06	1.30 ± 0.01	5.22±0.02	43.65 ± 0.04	17.45 ± 0.04	2.03 ± 0.01	[49]
REED	18.18	0.40	6.56	60.25	13.03	1.40	[51]
ANTILLANA	18.87	0.59	4.16	63.07	11.83	1.32	[51]
VARIETY NO	18.74 ± 0.06	0.51 ± 0.00	7.88 ± 0.01	54.40 ± 0.10	10.87 ± 0.01	0.61 ± 0.00	[3]
INDICATED	12.87	1.45	3.86	57.44	18.70	0.92	[53]

[1] composition studied under different extraction methods (cold pressed and solvents); [2] composition studied under different extraction methods (solvents, SCO_2, and UAAE); [3] in conditions similar to natural ripening.

In order to study the fatty acid profile of the avocado oil from the fruits harvested and artificially ripened, the avocado oil was extracted from preserved fruit at 5 °C in a controlled atmosphere and was analyzed (4% of O_2 and 6% of CO_2). Postharvest ripening was stimulated in the presence of exogenous ethylene (0 or 100 ppm) at a temperature of 18 °C for 24 h and then preserved at 15 and 20 °C. It was concluded that post-harvest conservation and ripening by means of a controlled atmosphere did not have a detrimental effect on the fatty acid profile or the amount of oil obtained, when compared with the one commonly applied in the ready-to-eat market [53]. This information is very important for the fresh product industry, where avocados move long distances at low temperatures.

When carrying out the physicochemical evaluation of two Hass avocado oils sold in Chile and labeled as extra virgin, it was demonstrated that both oils presented significant differences in the content of tocopherols, total phenols, oil stability, measured as the induction time, UV absorption coefficients, peroxide index, free acidity, total chlorophyll, total carotenoids, and polar compounds. In addition, the presence of 3,5-stigmastadiene in one of the samples, a compound that has been associated with a high degree of refining or exposure to high temperatures of oils, indicated a disparity in the quality parameters and a lack of regulation of avocado oil in the local market [37].

On the other hand, monovarietal oils from the *Bacon, Fuerte, Hass,* and *Pinkerton* varieties, obtained in Spain, were compared with commercial oils originating in Brazil, Chile, Ecuador and New Zealand. The content of triacylglycerols, fatty acids, aliphatic, and terpene alcohols, desmethylmethyl, methyl and dimethyl sterols, squalene and tocopherols were determined. The main triacylglycerols were those with ECN48 (48 equivalents of carbon atoms). The oleic, palmitic, and linoleic fatty acids were the most abundant fatty acids, and the desmethyl sterols were the main quantified minor compounds. Small amounts of aliphatic and terpene alcohols were observed. The concentrations of squalene were higher in the oils of the *Bacon, Fuerte,* and *Pinkerton* varieties than in the other varieties. The most abundant tocopherol was α-tocopherol [52].

While the fruit reaches a minimum of 8% fat at the time of its extraction from the tree, during vegetative ripening, values of 20% or more are reached, depending on the variety. In this sense, it is recommended that the avocado oil industry ripen the fruit on the tree, since climacteric ripening, in comparison with commercial ripening, influences not only the increase in oil content, but also the profile of the fatty acids, increasing the amount of unsaturated fatty acids, such as oleic acid, and decreasing the amount of saturated fatty acids, such as palmitic and palmitoleic acids [56]. This situation seems controversial with respect to the previously discussed studies. However, it is well known that the composition of fruits depends on environmental and growth conditions. In addition, the quantification of the different analytes depends on the conditions of extraction, processing, and detection limits of the analytical equipment.

Martinez-Nieto and Moreno-Romero [57] have analyzed the sterol composition of the unsaponifiable matter of avocado oil in 4 varieties (*Reed, Bacon, Fuerte,* and *Hass*) by means of gas chromatography. The results showed that the *Fuerte* and *Reed* varieties contained 2% more of β-sitosterol than the *Bacon* and *Hass* varieties. In relation to the cholesterol content, the smallest amount was present in the *Fuerte* variety and the largest in the Hass variety. In this sense, the unsaponifiable sterol composition can be a guiding tool to determine the authenticity of oil.

Martínez-Nieto et al. [51] analyzed the composition of different fractions in an industrial extraction process using a continuous method, under conditions similar to those used in the extraction of olive oil, for ripe avocados of the *Fuerte, Reed, Hass,* and *Antillana* varieties. They concluded that it is possible to obtain a good net oil yield with the industrial equipment used prior to specific modifications in the grinding stage and in the decanter, as well as good quality parameters, such as the acidity index, peroxide index, and absorbance coefficients, for both virgin oil and refined avocado oil.

When evaluating the physicochemical characteristics of the avocado oil obtained from the *Bantul, Purwokerto,* and *Garut* varieties, originating in Indonesia, by extraction with solvents, the *Garut* variety presented a better quality in terms of the iodine index, which is associated with a greater amount of unsaturated fatty acids. The conjugated dienes and trienes were significantly different between the samples. The p-anisidine index did not have significant differences between the samples. The saponification index was higher in the *Purwokerto* variety, and the peroxide index was higher in the *Bantul* variety. In addition, the analysis using differential scanning calorimetry (DSC) showed that the three samples had different melting and crystallization profiles [58].

6.3. Avocado Seed Oil

In relation to avocado seed oil, Barrera-López and Arrubia-Vélez [59] pointed out that the Lorena variety contained about 8.47% oil, and the unsaponifiable matter was 76.9%. The phytosterols quantified in a greater proportion were ergosterol, 5α-cholestane and stigmasterol.

Avocado seed presents, in its composition, a large number of extractable polyphenols, which have attracted attention due to their high antioxidant capacity. It was determined that, with a higher power of the ultrasound (0–104 W) and increase of the temperature (20–60 °C), the polyphenol content and antioxidant capacity was increased [60].

When performing a physicochemical analysis of the seed oil of the Hass variety, cultivated in Peru and obtained by the Soxhlet method, it was found that it had a high fatty acid profile in linoleic acid (48.77%) and linolenic acid (12.17%). While the antioxidant activity, determined by the DPPH method, was low, it was higher in the saponifiable fraction than in the unsaponifiable fraction, which was attributable to the presence of polyphenols and steroids. In addition, it was determined that the quality parameters, such as acidity, peroxide, saponification, iodine, and specific gravity indexes, were similar to those for extra virgin olive oil [61].

When comparing the composition of oil from the pulp and seed of the *Fuerte* variety, cultivated in the region of Northeastern Brazil, a great difference in the lipid content between the pulp and the seed can be seen (15.39% v.s. 1.87%, dry base). It was determined that the parameters of oil quality, refractive index, gravity, and peroxide index were similar for both oils, but the iodine, acid index, and saponification index were higher in seed oil than in pulp oil. Gas chromatography showed that seed oil had a greater variety of fatty acids than pulp oil. Additionally, the fatty acid profile of the pulp was much more concentrated in monounsaturated fatty acids than that of seed, and conversely, the seed oil is much more concentrated in polyunsaturated fatty acids than pulp oil [55].

6.4. Comparison with Other Oils

Dubois et al. (2007) [62] compared 80 varieties of vegetable oils, including avocado oil, indicating that it was composed of more than 60% of monounsaturated fatty acids, a characteristic shared with olive oil, hazelnut, and macadamia nut profiles. In comparison with olive oil, avocado oil possessed a higher proportion of saturated fatty acids (16.4%), with a predominance of palmitic acid (15.7%), a lower proportion of monounsaturated fatty acids (67.8%), with a predominance of oleic acid 60.3% and a higher proportion of polyunsaturated fatty acids (15.2%), the most important of which was linoleic acid at 13.7%.

Additionally, similar fatty acid profiles have been published by Berasategi et al. [3], who also showed that avocado oil had a higher PUFA/SFA and higher omega-6/omega-3 ratios than olive oil.

Berasategi et al. [3] showed that the phytosterol content was higher in avocado oil (3.3 g to 4.5 mg/g of oil) than in olive oil, of which the most abundant was β-sitosterol, followed by sitostanol, cycloartenol, cycloeucalenol and D7-avenasterol. The number of sterols in the avocado oil was higher, 4-demethyl-sterols being the most abundant, reaching 80% of the total fraction of sterols. This greater proportion was even maintained under drastic conditions of deterioration (180 °C). Moreover, avocado oil has a lower proportion of vitamin E, compared to olive oil. This study indicates that the thermal stability of avocado oil is similar to that of olive oil. From Table 2, it is possible to appreciate a summary of the different antioxidant components present in the avocado oils of the different varieties shown in this work.

Table 2. Antioxidant compounds present in avocado oil. Concentration [=] mg × kg^{-1}.

Varieties	β-Sitosterol	α-Tocopherol	γ-Tocopherol	Δ5-avenasterol	Campesterol	Estigmasterol	Sitoestanol	Campestanol	Ref.
HASS	91.917 ± 0.027	-	-	-	6.091 ± 0.026	0.001 ± 0.001	-	-	[57]
	95.2	-	-	-	4.7	0.13 ± 0.00	-	-	[51]
	82.95 ± 0.06	86.75 ± 0.62	9.02 ± 0.09	6.63 ± 0.07	5.88 ± 0.01	-	0.46 ± 0.01	0.04 ± 0.00	[52]
FUERTE	94.767 ± 0.012	-	-	-	5.043 ± 0.012	0.001 ± 0.001	0.57 ± 0.04	-	[57]
	92.9	-	-	-	6.4	-	-	-	[51]
BACON	80.56 ± 0.08	103.11 ± 6.87	20.35 ± 1.22	8.81 ± 0.03	4.62 ± 0.02	0.15 ± 0.00	-	0.04 ± 0.00	[52]
	92.189 ± 0.012	-	-	-	6.096 ± 0.010	0.011 ± 0.001	-	-	[57]
REED	82.6 ± 0.03	51.90 ± 0.04	71.61 ± 0.57	9.16 ± 0.03	3.71 ± 0.01	0.40 ± 0.01	0.58 ± 0.04	0.05 ± 0.00	[52]
ANTILLANA	94.605 ± 0.027	-	-	-	5.123 ± 0.021	0.001 ± 0.001	-	-	[57]
	91.2	-	-	-	8.6	-	-	-	[51]
PINKERTON	89.3	-	-	-	10.6	-	-	-	[51]
	84.08 ± 0.08	45.62 ± 0.19	13.71 ± 0.56	5.86 ± 0.01	6.00 ± 0.01	0.11 ± 0.00	0.41 ± 0.03	0.04 ± 0.00	[52]
VARIETY NO INDICATED	-	-	-	9.42 ± 1.69	18.36 ± 1.44	1.11 ± 0.12	2.19 ± 0.22	0.43 ± 0.03	[3]

7. Biological Effects

The presence of compounds with nutritional interest, such as unsaturated fatty acids (MUFA and PUFA), as well as compounds with biological activity, such as tocopherols, tocotrienols, phytosterols, carotenoids, and polyphenols, have made avocado oil of growing interest for research on the possible biological effects of avocado oil, with the aim of preventing and treating diseases through the diet of the population.

7.1. Human Health Effects

A study in 13 healthy adults with a habitual hypercaloric and hyperlipidic diet, where butter was replaced by avocado oil extracted at 35 °C from the pulp alone, was conducted. The incorporation of avocado oil for a period of six days reflected an improvement in the postprandial profile of insulin, glycemia, total cholesterol, low-density lipoproteins, triglycerides, and inflammatory parameters, such as C-reactive protein (CRP) and interleukin-6 [63]. Avocado pulp oil (Mexican creole genotypes) has shown anti-inflammatory activity by inhibiting the enzymes COX 1 and COX 2 in a similar way to the drug, ibuprofen, and extra virgin olive oil [48]. Additionally, when avocado oil was added to vitamin B12 skin cream preparation, it was well tolerated and had the potential for long-term topical therapy of psoriasis [64].

7.2. Experimental Studies in Animals

When administering avocado oil to Wistar diabetic rats for 90 days (1 gr/250 g weight), it was observed that it promoted an improvement in the functionality of the electron transport chain and decreased the generation of free radicals in the liver, attenuating the harmful effects of oxidative stress [65].

In the brain, an improvement in mitochondrial function has been observed, as well as a decrease in free radical levels, lipid peroxidation, and an improvement in the reduced/oxidized glutathione ratio. These results demonstrate that supplementation with avocado oil prevents mitochondrial dysfunction in the brain and liver of diabetic rats [66].

Ortiz-Avila et al. [67] demonstrated that supplementation with avocado oil in Wistar rats led to a reduction of the alterations in the electron transport chain at the renal level, attenuating the oxidative damage, although a protective effect on lipid peroxidation was not evidenced.

Some authors [68–70] studied the relationship between avocado oil and collagen metabolism in both the skin and liver, finding that oil obtained from intact fruit (pulp and seed), refined with hexane, was associated with fibrosis in the liver, an increase in liver enzymes and consequently hepatotoxicity.

In addition to an increase in the solubility of collagen, behavior that is attributed to a decrease in the activity of lysyl oxidase was observed [66]. A similar conclusion was proposed by Lamaud et al. [71], stating that the mixture of soybean and avocado oils decreased the degree of collagen cross-linking, a process that has been associated with a delay in wound healing. However, Oliveira et al. [53] indicated that the healing activity of a semi-solid formulation of avocado oil (50/50 Vaseline) and avocado pulp oil promoted the increase of collagen synthesis and decreased the number of inflammatory cells during the process of wound healing.

Regarding the effects on cardiovascular health, it is possible to point out that the atherogenic power of avocado oil could be similar to that of olive oil and lower than that of corn and coconut oil [72]. On the other hand, an avocado oil-rich diet modifies the fatty acid content in cardiac and renal membranes in a tissue-specific manner in Wistar rats. The rise in renal arachidonic acid suggests that diet content can be a key factor in vascular responses [73].

Márquez-Ramírez et al. [74] studied the antihypertensive effect of avocado oil in rats with induced hypertension. They were subsequently treated with the administration of 1 mL of oil per 250 g of rat weight or 40 mg of losartan potassium per kg of rat weight. Avocado oil significantly decreased both systolic and diastolic pressure in hypertensive rats, but not in controls. Avocado oil mimicked the effects of the drug, losartan, on blood pressure, vascular performance, and oxidative stress.

In rats fed with sucrose, it was observed that the addition of avocado oil in the diet reduced the levels of triglycerides, VLDL and LDL, without affecting the levels of HDL. It also reduced the level of ultrasensitive CPR, indicating that the inflammatory processes associated with metabolic syndrome were partially re-established [4]. Additionally, a diet high in sucrose, which causes liver alteration in Sprague-Dawley Weaned rats, was partially reversible by the administration of avocado oil, obtained by centrifugation and extracted by solvents. This effect was similar to that of olive oil, but would not be extended to the pancreatic level [5]. In relation to insulin resistance induced by a diet high in sucrose in Wistar rats, it was possible to reduce it through the dietary addition of 5–20% avocado oil [75]. In female Wistar rats exposed to prolonged androgenic stimulation, avocado oil improved triglyceride, VLDL and HDL levels, generating a direct regulatory effect on the lipid profile [76]. The influence of avocado oil on cell damage, induced by methyl methanesulfonate (MMS) and doxorubicin, can be carried out using in vitro (V79 cells) and in vivo models (Swiss mice). Avocado pulp oil had no genotoxic effects. The oil was effective in reducing the chromosomal damage induced by MMS and doxorubicin. However, an increase in liver tissue damage was observed when evaluating a high dose of avocado oil (measured as an increase in the hepatic enzyme aspartate aminotransferase), a phenomenon attributable to the high concentration of palmitic fatty acid [77]. Table 3 shows a summary of the biological effects presented in this study.

Table 3. Experimental studies on the biological effects of avocado.

Animal Model	Protocol Used	Conclusions	Ref.
Male Wistar rats	Daily administration of 1.0 mL/250 g of avocado oil by gavage or losartan at 40 mg/kg for 45 days.	(a) Avocado oil mimics the effects of losartan. (b) Effects of avocado oil could be mediated by decreased actions of Angiotensin-II in mitochondria. (c) The intake of avocado oil could mitigate the harmful effects of hypertension in kidneys.	[74]
Diabetic male rats Goto-Kakizaki; the control rats were Wistar males.	Daily administration of avocado oil was 1 mL/250 g of weight for a period of 3, 6 and 12 months to Winstar and Goto-Kakizaki rats.	In the brain: (a) Improvement of mitochondrial function. (b) Decreased levels of free radicals and lipid peroxidation. (c) Improvement of the reduced/oxidized glutathione ratio. (d) Prevention of mitochondrial dysfunction.	[66]
In vitro (Chinese hamster lung fibroblasts/V79 cells) and in vivo models (Swiss mice).	Swiss mice were given avocado oil in different concentrations of 250, 500, 1000 and 2000 mg/kg in an in vivo model. In vitro model was applied to different concentrations of avocado oil: 100, 200 and 400 µg/mL.	(a) Avocado pulp oil has no genotoxic effects. (b) The oil was effective in reducing the chromosomal damage induced by methyl methanesulfonate and doxorubicin. However, an increase in hepatic enzyme aspartate aminotransferase was found to be a marker of liver damage.	[77]
Wistar rats	Administration of a diet with different concentrations of avocado oil to insulin resistant rats.	(a) The dietary addition of 5–20% avocado oil can reduce glucose tolerance and insulin resistance, induced by the high sucrose diet, in Wistar rats. (b) The addition of 5–30% avocado oil reduced the body weight gain induced by the high sucrose diet in Wistar rats.	[75]
Male Wistar rats	Administration of avocado oil (1 g/250 g weight) to diabetic rats for 90 days.	In the liver: (a) Promoted an improvement in the functionality of the electron transport chain. (b) Decreased the generation of free radicals. (c) Diminished the harmful effects of oxidative stress in the liver.	[65]
Male Wistar rats	Diet supplemented with avocado oil to rats exposed to prolonged androgenic stimulation.	Avocado oil exerted a direct regulatory effect on the lipid profile.	[76]
Wistar rats	Administration of a diet with olive oil and avocado oil to rats exposed to a diet rich in sucrose.	In the liver: (a) A diet high in sucrose causes hepatic impairment, partially reversible by the administration of avocado oil, which does not occur at the pancreatic level. (b) Avocado oil (independent of its extraction method) exhibits effects similar to those of olive oil in the fatty acid profile.	[5]
Wistar rats	Administration of a diet with olive oil and avocado oil to rats exposed to a diet rich in sucrose.	At the cardiovascular level: In rats fed with sucrose, it was observed that avocado oil reduces the levels of triglycerides, very low-density lipoprotein (VLDL) and low-density lipoprotein (LDL), without affecting the levels of high-density lipoprotein (HDL). It also reduces the level of ultrasensitive CRP, indicating that the inflammatory processes associated with metabolic syndrome are partially reestablished.	[4]

Table 3. *Cont.*

Animal Model	Protocol Used	Conclusions	Ref.
Male Wistar rats	Avocado oil administration: 1 mL of avocado oil/250 g of weight daily for a period of 90 days.	At the renal level: (a) Protection against induced oxidative stress. (b) ROS generation decrease. (c) Improvement of the activities of complexes II and III. (d) Lower peroxidizability index in diabetic mitocondria.	[67]
Male Wistar rats	The control group received a laboratory pellet, while the treated group received a diet rich in 10% avocado oil (weight/weight) for a period of two weeks.	The administration of a diet rich in avocado oil to rats for two weeks modifies the content of fatty acids in cardiac and renal membranes.	[73]
Male rabbits of the New Zealand White strain.	Rabbits were fed a semi-purified diet containing 0.2% cholesterol and 14% oil for 90 days (corn, coconut, olive and avocado).	The atherogenic power of avocado oil is equivalent to that of olive oil and lower than that of corn and coconut oil.	[72]
Female rats and chicks.	Rats and chicks were fed 10% avocado oil (*w/w*).	Rats and chickens fed unrefined avocado oil showed a significant decrease in total collagen solubility in the liver.	[68]
Charles river female rats.	Growing rats were fed for eight weeks with 10% (*w/w*) refined or unrefined avocado oil or soybean oil, using different extraction media.	Rats fed with unrefined avocado oil extracted with hexane from intact fruit (unsaponifiables) or avocado seed oil showed significant increases in the content of soluble collagen in the skin, although the total collagen content was not affected. Rats fed refined or unrefined soybean oils showed no effects.	[70]
Charles river female rats.	The rats were fed diets containing 10% avocado oil (*w/w*) for four weeks.	The consumption of avocado oil extracted from intact fruits can cause changes in metabolism in the liver. Serum alkaline phosphatase activity increased in rats fed seed oil, oil extracted with unrefined solvent of intact fruit or unsaponifiables, and aspartate aminotransferase activity increased significantly in the group fed avocado seed oil.	[69]
Rats	Application of oil mixtures in the skin of the dorsal area of rats for 15 days, treatment with a mixture of 2/3 of soybean oil and 1/3 avocado in a 5% solution of sweet almond oil.	The mixture of soybean and avocado oils decreases the degree of collagen cross-linking, delaying wound healing.	[71]

8. Conclusions

While the composition and quality of the avocado oil depends on the origin, weather conditions, variety, and extraction methods, it is characterized as a mainly monounsaturated oil, with an adequate proportion of polyunsaturated fatty acids, similar to olive oil. In addition, it contains other bioactive compounds, present in the unsaponifiable fraction, such as tocopherols, polyphenols, and a remarkable proportion of phytosterols. This oil has also been shown to perform well at high temperatures. All these characteristics indicate that avocado oil has nutritional properties that are greatly appreciated by the population, even for technological applications. This makes it a product of increasing exploitation interest for producers.

This growing interest also requires a greater development of studies on the adulteration and contamination of avocado oil, as well as the further assessment of the promising biological effects of the different components present in the oil, either in animals or humans. Furthermore, considering the presented information, it is necessary to inquire into the international regulations concerning the different qualities of this oil.

Author Contributions: Conceptualization, M.F. and J.O.-V.; methodology, C.E.V., C.S. and J.O.-V.; software, H.V.; validation, C.E.V., F.A., and H.V.; formal analysis, M.F. and F.A.; investigation, M.F., C.S. and J.O.-V.; resources, M.F.; data curation, C.S. and C.E.V.; writing—original draft preparation, C.S.; writing—review and editing, M.F., F.A., H.V. and J.O.-V.

Funding: This research received no external funding.

Abbreviations

UAAE	Ultrasound-assisted aqueous extraction
sCO$_2$	Subcritical CO$_2$
scCO$_2$	Supercritical CO$_2$
LPG	Liquefied gas of compressed oil
FFA	Free fatty acid
MUFA	Monounsaturated fatty acid
PUFA	Polyunsaturated fatty acid
SFA	Saturated fatty acid
TAG	Triacylglycerol
O	Oleic acid
P	Palmitic acid
L	Linoleic acid
COX	Cyclooxygenase enzyme
PCR	C reactive protein
FTIR	Infrared spectroscopy with Fourier transform
IOC	International olive council
DPPH	2,2-Diphenyl-1-picrylhydrazyl
APMAE	Atmospheric pressure microwave-assisted liquid–liquid extraction
PHAs	Polyhydroxyalkanoates
ECN48	48 equivalents of carbon atoms
VLDL	Very low-density lipoprotein
LDL	Low-density lipoprotein
HDL	High-density lipoprotein

References

1. Tan, C.; Tan, S.; Tan, S. Influence of Geographical Origins on the Physicochemical Properties of *Hass* Avocado Oil. *J. Am. Oil Chem. Soc.* **2017**, *94*, 1431–1437. [CrossRef]

2.	Wong, M.; Requejo-Jackman, C.; Woolf, A. What is unrefined, extra virgin cold-pressed avocado oil? *Inform* **2010**, *21*, 189–260.

3.	Berasategi, I.; Barriuso, B.; Ansorena, D.; Astiasarán, I. Stability of avocado oil during heating: Comparative study to olive oil. *Food Chem.* **2012**, *1*, 439–446. [CrossRef] [PubMed]

4.	Carvajal-Zarrabal, O.; Nolasco-Hipolito, C.; Aguilar-Uscanga, M.; Melo-Santiesteban, G.; Hayward-Jones, P.; Barradas-Dermitz, D. Avocado oil supplementation modifies cardiovascular risk profile markers in a rat model of sucrose-induced metabolic changes. *Dis. Markers* **2014**, *2014*, 386–425. [CrossRef] [PubMed]

5.	Carvajal-Zarrabal, O.; Nolasco-Hipolito, C.; Aguilar-Uscanga, M.; Melo-Santiesteban, G.; Hayward-Jones, P.; Barradas-Dermitz, D. Effect of dietary intake of avocado oil and olive oil on biochemical markers of liver function in sucrose-fed rats. *Biomed. Res. Int.* **2014**, *2014*, 595479. [CrossRef] [PubMed]

6.	Di Stefano, V.; Avellone, G.; Bongiorno, D.; Indelicato, S.; Massenti, R.; Lo Bianco, R. Quantitative evaluation of the phenolic profile in fruits of six avocado (*Persea americana*) cultivars by ultra-high-performance liquid chromatography-heated electrospray-mass spectrometry. *Int. J. Food Prop.* **2017**, *20*, 1302–1312. [CrossRef]

7.	Woolf, A.; Wong, M.; Eyres, L.; McGhie, T.; Lund, C.; Olsson, S.; Wang, Y.; Bulley, C.; Wang, M.; Friel, E.; et al. Avocado oil. From cosmetic to culinary oil. In *Gourmet and Health-Promoting Specialty Oils*; Moreau, R., Kamal-Eldin, A., Eds.; AOCS Press: Urbana, IL, USA, 2009; pp. 73–125.

8.	NM. 2008. Norma Mexicana NMX-F-052- SCFI-**2008**. Aceites y grasas-aceite de aguacate - especificaciones. Available online: http://aniame.com/mx/wp-content/uploads/Normatividad/CTNNIAGS/NMX-F-052-SCFI-2008.pdf (accessed on 7 June 2019).

9.	Costagli, G.; Betti, M. Avocado oil extraction processes: Method for cold-pressed high-quality edible oil production versus traditional production. *J. Agric. Eng.* **2015**, *46*, 115–122. [CrossRef]

10.	Krumreich, F.D.; Borges, C.D.; Mendonça, C.R.B.; Jansen-Alves, C.; Zambiazi, R.C. Bioactive compounds and quality parameters of avocado oil obtained by different processes. *Food Chem.* **2018**, *257*, 376–381. [CrossRef]

11.	Santana, I.; dos Reis, L.; Torres, A.; Cabral, L.; Freitas, S. Avocado (*Persea americana Mill.*) oil produced by microwave drying and expeller pressing exhibits low acidity and high oxidative stability. *Eur. J. Lipid Sci. Technol.* **2015**, *117*, 999–1007. [CrossRef]

12.	Chimsook, T.; Assawarachan, R. Effect of Drying Methods on Yield and Quality of the Avocado Oil. *Adv. Mater. Res. Key Eng. Mater.* **2017**, *735*, 127–131. [CrossRef]

13.	Dos Santos, M.; Alicieo, T.V.; Pereira, C.M.; Ramis-Ramos, G.; Mendonça, C.R. Profile of bioactive compounds in avocado pulp oil: Influence of the drying processes and extraction methods. *J. Am. Oil Chem. Soc.* **2014**, *91*, 19–27. [CrossRef]

14.	Norma para grasas y aceites comestibles no regulados por normas individuales. Available online: http://URLwww.fao.org/input/download/standards/74/CXS_019s_2015.pdf (accessed on 16 April 2019).

15.	Yang, S.; Hallett, I.; Rebstock, R.; Oh, H.; Kam, R.; Woolf, A.; Wong, M. Cellular Changes in "*Hass*" Avocado Mesocarp During Cold-Pressed Oil Extraction. *J. Am. Oil Chem. Soc.* **2018**, *95*, 229–238. [CrossRef]

16.	Xuan, T.; Hean, C.; Hamzah, H.; Ghazali, H. Optimization of ultrasound-assisted aqueous extraction to produce virgin avocado oil with low free fatty acids. *J. Food Process Eng.* **2017**, *12656*, 1–9.

17.	Martínez-Padilla, L.P.; Franke, L.; Xu, X.Q.; Juliano, P. Improved extraction of avocado oil by application of sono-physical processes. *Ultrason. Sonochemistry* **2018**, *40*, 720–726. [CrossRef] [PubMed]

18.	Valdés, H.; Unda, K.; Saavedra, A. Numerical Simulation on Supercritical CO_2 Fluid Dynamics in a Hollow Fiber Membrane Contactor. *Computation* **2019**, *7*, 8. [CrossRef]

19.	Corzzini, S.; Barros, H.; Grimaldi, R.; Cabral, F. Extraction of edible avocado oil using supercritical CO_2 and a CO_2/ethanol mixture as solvents. *J. Food Eng.* **2017**, *194*, 40–45. [CrossRef]

20.	Barros, H.D.; Coutinho, J.P.; Grimaldi, R.; Godoy, H.T.; Cabral, F.A. Simultaneous extraction of edible oil from avocado and capsanthin from red bell pepper using supercritical carbon dioxide as solvent. *J. Supercrit. Fluids* **2016**, *107*, 315–320. [CrossRef]

21.	Barros, H.; Grimaldi, R.; Cabral, F. Lycopene-rich avocado oil obtained by simultaneous supercritical extraction from avocado pulp and tomato pomace. *J. Supercrit. Fluids* **2017**, *120*, 1–6. [CrossRef]

22.	Restrepo, A.M.; Londoño, J.; González, D.; Benavides, Y.; Cardona, B.L. Comparación del aceite de aguacate variedad *Hass* cultivado en Colombia, obtenido por fluidos supercríticos y métodos convencionales: Una perspectiva desde la calidad. *Rev. Lasallista Investig.* **2012**, *9*, 151–161.

23. Abaide, E.; Zabot, G.; Tres, M.; Martins, R.; Fagundez, J.; Nunes, L.; Druzian, S.; Soares, J.; Dal Prá, V.; Silva, J.; et al. Yield composition, and antioxidant activity of avocado pulp oil extracted by pressurized fluids. *Food Bioprod. Process.* **2017**, *102*, 289–298. [CrossRef]

24. Xuan, T.; Hean, C.; Hamzah, H.; Ghazali, H. Comparison of subcritical CO_2 and ultrasound-assisted aqueous methods with the conventional solvent method in the extraction of avocado oil. *J. Supercrit. Fluids* **2018**, *135*, 45–51.

25. AOAC. *Official Methods of Analysis of AOAC International*, 18th ed.; Horwitz, W., Latimer, G., Eds.; AOAC International: Maryland, MD, USA, 2007.

26. Buenrostro, M.; López-Munguia, A.C. Enzymatic extraction of avocado oil. *Biotechnol. Lett.* **1986**, *8*, 505–506. [CrossRef]

27. Reddy, M.; Moodley, R.; Jonnalagadda, S.B. Fatty acid profile and elemental content of avocado (*Persea americana Mill.*) oil-effect of extraction methods. *J. Environ. Sci. Health B* **2012**, *47*, 529–537. [CrossRef] [PubMed]

28. Meyer, M.D.; Terry, L.A. Development of a rapid method for the sequential extraction and subsequent quantification of fatty acids and sugars from avocado mesocarp tissue. *J. Agric. Food Chem.* **2008**, *56*, 7439–7445. [CrossRef] [PubMed]

29. Lewis, C.E.; Morris, R.; O'Brien, K. The oil content of avocado mesocarp. *J. Sci. Food Agric.* **1978**, *29*, 943–949. [CrossRef] [PubMed]

30. Ortiz-Moreno, A.; Dorantes, L.; Galíndez, J.; Guzmán, R.I. Effect of different extraction methods on fatty acids, volatile compounds, and physical and chemical properties of avocado (*Persea americana Mill.*) oil. *J. Agric. Food Chem.* **2003**, *51*, 2216–2221. [CrossRef] [PubMed]

31. Mostert, M.E.; Botha, B.M.; Du Plessis, L.M.; Duodu, K.G. Effect of fruit ripeness and method of fruit drying on the extractability of avocado oil with hexane and supercritical carbon dioxide. *J. Sci. Food Agric.* **2007**, *87*, 2880–2885. [CrossRef]

32. Bizimana, V.; Breene, W.M.; Csallany, A.S. Avocado oil extraction with appropriate technology for developing countries. *J. Am. Oil Chem. Soc.* **1993**, *70*, 821–822. [CrossRef]

33. Ariza-Ortega, J.A.; Cruz-Cansino, N.; Ramírez-Moreno, E.; Ramos-Cassellis, M.E.; Castañeda-Antonio, D.; Betanzos-Cabrera, G. Effect of electric field on the characteristics of crude avocado oil and virgin olive. *J. Food Sci. Technol.* **2017**, *54*, 2166–2170. [CrossRef] [PubMed]

34. Ariza-Ortega, J.; Mendez-Ramos, M.; Diaz-Reyes, J.; Delgado-Macuil, R.; De-la-Torre, R. Study by Fourier Transform Infrared Spectroscopy of the Avocado Oils of the Varieties *Hass, Criollo* and *Fuerte*. *J. Mater. Sci. Eng.* **2010**, *4*, 61–64.

35. Cerecedo-Cruz, L.; Azuara-Nieto, E.; Hernández-Álvarez, A.J.; González-González, C.R.; Melgar-Lalanne, G. Evaluation of the oxidative stability of Chipotle chili (Capsicum annuum L.) oleoresins in avocado oil. *Grasas Aceites* **2018**, *69*, 240. [CrossRef]

36. Cicero, N.; Albergamo, A.; Salvo, A.; Bua, G.; Bartolomeo, G.; Mangano, V.; Rotondo, A.; Di Stefano, V.; Di Bella, G.; Dugo, G. Chemical characterization of a variety of cold-pressed gourmet oils available on the Brazilian market. *Food Res. Int.* **2018**, *109*, 517–525. [CrossRef] [PubMed]

37. Retief, L.; McKenzieb, JM.; Koch, K. A novel approach to the rapid assignment of 13C NMR spectra of major components of vegetable oils such as avocado, mango kernel and macadamia nut oils. *Magn. Reson. Chem.* **2009**, *47*, 771–781. [CrossRef] [PubMed]

38. Fernándes, G.D.; Gómez-Coca, R.B.; Pérez-Camino, M.C.; Moreda, W.; Barrera-Arellano, D. Chemical characterization of commercial and single-variety avocado oils. *Grasas Aceites* **2018**, *69*, 256. [CrossRef]

39. Castejón, D.; Mateos-Aparicio, I.; Molero, M.D.; Cambero, MI.; Herrera, A. Evaluation and Optimization of the Analysis of Fatty Acid Types in Edible Oils by 1 H-NMR. *Food Anal. Methods* **2014**, *7*, 1285–1297. [CrossRef]

40. Rohman, A.; Windarsih, A.; Riyanto, S.; Sudjadi, S.A.S.A.; Rosman, A.S.; Yusoff, F.M. Fourier transform infrared spectroscopy combined with multivariate calibrations for the authentication of avocado oil. *Int. J. Food Prop.* **2016**, *19*, 680–687. [CrossRef]

41. Rohman, A.; Lumakso, F.A.; Riyanto, S. Use of partial least square-discriminant analysis combined with mid infrared spectroscopy for avocado oil authentication. *J. Med. Plants Res.* **2016**, *10*, 175–180. [CrossRef]

42. Fuentes, E.; Báez, M.E.; Díaz, J. Microwave-assisted extraction at atmospheric pressure coupled to different clean-up methods for the determination of organophosphorus pesticides in olive and avocado oil. *J. Chromatogr. A* **2009**, *1216*, 8859–8866. [CrossRef]

43. Arancibia, C.; Riquelme, N.; Zúñiga, R.; Matiacevicha, S. Comparing the effectiveness of natural and synthetic emulsifiers on oxidative and physical stability of avocado oil-based nanoemulsions. *Innov. Food Sci. Emerg. Technol.* **2017**, *44*, 159–166. [CrossRef]

44. Caballero, E.; Soto, C.; Olivares, A.; Altamirano, C. Potential use of avocado oil on structured lipids MLM-type production catalysed by commercial immobilised lipases. *PloS ONE* **2014**, *9*, E107749. [CrossRef]

45. Flores-Sánchez, A.; López-Cuellar, M.; Pérez-Guevara, F.; Figueroa López, U.; Martín-Bufájer, J.M.; Vergara-Porras, B. Synthesis of Poly-(R-hydroxyalkanoates) by Cupriavidus necator ATCC 17699 Using Mexican Avocado (*Persea americana*) Oil as a Carbon Source. *Int. J. Polym. Sci.* **2017**. [CrossRef]

46. Forero-Doria, O.; Flores, M.; Vergara, C.E.; Guzman, L. Thermal analysis and antioxidant activity of oil extracted from pulp of ripe avocados. *J. Therm. Anal. Calorim.* **2017**, *130*, 959–966. [CrossRef]

47. Galvão, M.D.S.; Narain, N.; Nigam, N. Influence of different cultivars on oil quality and chemical characteristics of avocado fruit. *Food Sci. Technol.* **2014**, *34*, 539–546. [CrossRef]

48. Espinosa-Alonso, L.G.; Paredes-López, O.; Valdez-Morales, M.; Oomah, B.D. Avocado oil characteristics of Mexican creole genotypes. *Eur. J. Lipid Sci. Technol.* **2017**, *119*, 1600406. [CrossRef]

49. Yanty, N.A.M.; Marikkar, J.M.N.; Long, K. Effect of varietal differences on composition and thermal characteristics of avocado oil. *J. Am. Oil Chem. Soc.* **2011**, *88*, 1997–2003. [CrossRef]

50. Dreher, M.; Davenport, A. *Hass* Avocado Composition and Potential Health Effects. *Crit. Rev. Food Sci. Nutr.* **2013**, *53*, 738–750. [CrossRef]

51. Martínez-Nieto, L.; Barranco-Barranco, R.; Moreno-Romero, M.V. Extracción de aceite de aguacate: Un experimento industrial. *Grasas Aceites* **1992**, *43*, 11–15. [CrossRef]

52. Flores, M.A.; Perez-Camino, M.C.; Troca, J. Preliminary Studies on Composition, Quality and Oxidative Stability of Commercial Avocado Oil Produced in Chile. *J. Food Sci. Eng.* **2014**, *4*, 21–26.

53. Pedreschi, R.; Hollak, S.; Harkema, H.; Otma, E.; Robledo, P.; Westra, E.; Defilippi, B.G. Impact of postharvest ripening strategies on '*Hass*' avocado fatty acid profiles. *South Afr. J. Bot.* **2016**, *103*, 32–35. [CrossRef]

54. Oliveira, A.; De Souza, E.; Rodrigues, R.; Cordeiro, D.; Gonçalves, R.; Ferreira, C.; Rodrigues, A.; De Medeiros, P.; Gonçalves, T.; Da Silva, A.; et al. Effect of Semisolid Formulation of *Persea Americana Mill* (avocado) Oil on Wound Healing in Rats. *Evid. -Based Complement. Altern. Med.* **2013**, *2013*, 1–8. [CrossRef]

55. Bora, P.S.; Narain, N.; Rocha, R.V.; Paulo, M.Q. Characterization of the oils from the pulp and seeds of avocado (cultivar: *Fuerte*) fruits. *Grasas Aceites* **2001**, *52*, 171–174.

56. Martinez-Nieto, L.; Moreno-Romero, M.V. Evolución del contenido de ácidos grasos de aceite de aguacate durante la maduración. *Grasas Aceites* **1995**, *46*, 92–95. [CrossRef]

57. Martinez-Nieto, L.; Moreno-Romero, M.V. Sterolic composition of the unsaponifiable fraction of oil of avocado of several varieties. *Grasas Aceites* **1994**, *45*, 402–403.

58. Indriyani, L.; Rohman, A.; Riyanto, S. Physico-chemical characterization of avocado (*Persea americana Mill.*) oil from three Indonesian avocado cultivars. *Res. J. Med. Plants* **2016**, *10*, 67–78.

59. Barrera-López, R.E.; Arrubla-Vélez, J.P. Análisis de Fitoesteroles en la Semilla de *Persea americana Miller* (var. Lorena) por Cromatografía de Gases y Cromatografía Líquida de Alta Eficiencia. *Rev. Fac. Cienc. Básicas* **2017**, *13*, 35–41.

60. Segovia, F.J.; Corral-Pérez, J.J.; Almajano, M.P. Avocado seed: Modeling extraction of bioactive compounds. *Ind. Crop. Prod.* **2016**, *85*, 213–220. [CrossRef]

61. Rengifo, P.G.; Carhuapoma, M.; Artica, L.; Castro, A.J.; López, S. Caracterización y actividad antioxidante del aceite de semilla de palta *Persea americana MILL. Cienc. E Investig.* **2015**, *18*, 33–36.

62. Dubois, V.; Breton, S.; Linder, M.; Fanni, J.; Parmentier, M. Fatty acid profiles of 80 vegetable oils with regard to their nutritional potential. *Eur. J. Lipid Sci. Technol.* **2007**, *109*, 710–732. [CrossRef]

63. Furlan, C.P.B.; Valle, S.C.; Östman, E.; Maróstica, M.R.; Tovar, J. Inclusion of *Hass* avocado-oil improves postprandial metabolic responses to a hypercaloric-hyperlipidic meal in overweight subjects. *J. Funct. Foods* **2017**, *38*, 349–354. [CrossRef]

64. Stucker, M.; Memmel, U.; Hoffmann, M.; Hartung, J.; Altmeyer, P. Vitamin B12 cream containing avocado oil in the therapy of plaque psoriasis. *Dermatology* **2001**, *203*, 141–147. [CrossRef]

65. Ortiz-Avila, O.; Gallegos-Corona, M.; Sánchez-Briones, L.; Calderón-Cortés, E.; Montoya-Pérez, R.; Rodriguez-Orozco, A.; Campos-García, J.; Saavedra-Molina, A.; Mejía-Zepeda, R.; Cortés-Rojo, C. Protective effects of dietary avocado oil on impaired electron transport chain function and exacerbated oxidative stress in liver mitochondria from diabetic rats. *J. Bioenerg. Biomembr.* **2015**, *47*, 337–353. [CrossRef] [PubMed]

66. Ortiz-Avila, O.; Esquivel-Martínez, M.; Olmos-Orizaba, B.; Saavedra-Molina, A.; Rodriguez-Orozco, A.; Cortés-Rojo, C. Avocado Oil Improves Mitochondrial Function and Decreases Oxidative Stress in Brain of Diabetic Rats. *J. Diabetes Res.* **2015**, *2015*, 485759. [CrossRef]

67. Ortiz-Ávila, O.; Sámano-García, C.; Calderón-Cortés, E.; Pérez-Hernández, I.; Mejía-Zepeda, R.; Rodríguez-Orozco, A.; Saavedra-Molina, A.; Cortés-Rojo, C. Dietary avocado oil supplementation attenuates the alterations induced by type I diabetes and oxidative stress in electron transfer at the complex II-complex III segment of the electron transport chain in rat kidney mitochondria. *J. Bioenerg. Biomembr.* **2013**, *45*, 271–287. [CrossRef] [PubMed]

68. Wermam, M.J.; Mokady, S.; Neeman, I. Effect of dietary avocado oils on hepatic collagen metabolism. *Ann. Nutr. Metab.* **1991**, *35*, 253–260. [CrossRef] [PubMed]

69. Werman, M.; Mokady, S.; Neeman, I.; Auslaender, L.; Zeidler, A. The effect of avocado oils on some liver characteristics in growing rats. *Food Chem. Toxicol.* **1989**, *27*, 279–282. [CrossRef]

70. Wermam, M.J.; Mokady, S.; Nimni, M.E.; Neeman, I. The effect of various avocado oils on skin collagen metabolism. *Connect. Tissue Res.* **1990**, *26*, 1–10. [CrossRef]

71. Lamaud, E.; Huc, A.; Wepierre, J. Effects of avocado and soya bean lipidic non-saponifiables on the components of skin connective tissue after topical application in the hairless rat: Biophysical and biomechanical determination. *Int. J. Cosmet. Sci.* **1982**, *4*, 143–152. [CrossRef]

72. Kritchevsky, D.; Tepper, S.; Wright, S.; Czarnecki, S.; Wilson, T.; Nicolosi, R. Cholesterol vehicle in experimental atherosclerosis 24: Avocado oil. *J. Am. Coll. Nutr.* **2003**, *22*, 52–55. [CrossRef]

73. Salazar, M.J.; El Hafidi, M.; Pastelin, G.; Ramírez-Ortega, M.C.; Sánchez-Mendoza, M.A. Effect of an avocado oil-rich diet over an angiotensin II-induced blood pressure response. *J. Ethnopharmacol.* **2005**, *98*, 335–338. [CrossRef]

74. Márquez-Ramírez, C.; Hernández de la Paz, J.; Ortiz-Avila, O.; Raya-Farias, A.; González-Hernández, J.; Rodríguez-Orozco, A.; Salgado-Garciglia, R.; Saavedra-Molina, A.; Godínez-Hernández, D.; Cortés-Rojo, C. Comparative effects of avocado oil and losartan on blood pressure, renal vascular function, and mitochondrial oxidative stress in hypertensive rats. *Nutrition* **2018**, *54*, 60–67.

75. Del Toro-Equihua, M.; Velasco-Rodríguez, R.; López-Ascencio, R.; Vásquez, C. Effect of an avocado oil-enhanced diet (*Persea americana*) on sucrose-induced insulin resistance in Wistar rats. *J. Food Drug Anal.* **2016**, *24*, 350–357. [CrossRef] [PubMed]

76. De Souza-Abboud, R.; Alves-Pereira, V.; Soares da Costa, C.A.; Teles-Boaventura, G.; Alves-Chagas, M. The action of avocado oil on the lipidogram of wistar rats submitted to prolonged androgenic stimulum. *Nutr. Hosp.* **2015**, *32*, 696–701.

77. Nicolella, H.D.; Neto, F.R.; Corrêa, M.B.; Lopes, D.H.; Rondon, E.N.; Dos Santos, L.F.; De Oliveira, P.F.; Damasceno, J.L.; Acésio, N.O.; Turatti, I.C.; et al. Toxicogenetic study of *Persea americana* fruit pulp oil and its effect on genomic instability. *Food Chem. Toxicol.* **2017**, *101*, 114–120. [CrossRef] [PubMed]

MDPI

St. Alban-Anlage 66

4052 Basel

Switzerland

Tel. +41 61 683 77 34

Fax +41 61 302 89 18

www.mdpi.com

Molecules Editorial Office

E-mail: molecules@mdpi.com

www.mdpi.com/journal/molecules